Tutorials, Schools, and Workshops in the Mathematical Sciences

Birkhäuser

The series is the primary resource for the publication of results and developments presented at summer or winter schools, workshops, tutorials, and seminars. Written in an informal and accessible style, they present important and emerging topics in scientific research for PhD students and researchers. Filling a gap between traditional lecture notes, proceedings, and standard textbooks, the titles included in TSWMS present material from the forefront of research.

More information about this series at http://www.springer.com/series/15641

Vicente Cortés • Klaus Kröncke • Jan Louis
Editors

Geometric Flows and the Geometry of Space-time

Birkhäuser

Editors
Vicente Cortés
Department of Mathematics
University of Hamburg
Hamburg, Germany

Klaus Kröncke
Department of Mathematics
University of Hamburg
Hamburg, Germany

Jan Louis
Department of Physics
University of Hamburg
Hamburg, Germany

ISSN 2522-0969 ISSN 2522-0977 (electronic)
Tutorials, Schools, and Workshops in the Mathematical Sciences
ISBN 978-3-030-01125-3 ISBN 978-3-030-01126-0 (eBook)
https://doi.org/10.1007/978-3-030-01126-0

Library of Congress Control Number: 2018960861

Mathematics Subject Classification (2010): 53C44, 53C50, 53C27, 53C29

This book is published under the imprint Birkhäuser, www.birkhauser-science.com by the registered company Springer Nature Switzerland AG
The registered company address is: Gewerbestrasse 11, 6330 Cham, Switzerland

Preface

A major goal in mathematics as well as in physics has been and still is to understand the geometry of space and time. Developments in both subjects have fruitfully influenced each other over the history of science. The formulation of general relativity by Einstein would not have been possible without the concepts of (semi-) Riemannian geometry that had emerged with the visionary ideas of Riemann in the previous century. Conversely, ideas from general relativity influenced mathematical research and the study of Einstein's equation is one of today's major topics in geometric analysis.

Similarly, the development of more recent areas of theoretical physics, such as string theory, is deeply connected to the study of geometric problems in mathematics, such as the study of metrics of special holonomy. It turned out that geometric flows are also of great importance in the interplay between mathematics and physics; e.g., the Riemann Penrose inequality has been shown by Huisken and Ilmanen using the inverse mean curvature flow.

This volume is based on a summer school and workshop entitled "Geometric flows and the geometry of space-time" held at the University of Hamburg in September 2016. The aim of this event was to provide a forum where physicists and mathematicians can exchange ideas and where graduate students and young researchers get the opportunity to learn about recent developments at the intersection of mathematics and physics.

It brought together around 60 participants with mathematical and physical backgrounds. The speakers were Lars Andersson, Helga Baum, Spiros Cotsakis, Pau Figueras, Gary Gibbons, Mark Haskins, Jason Lotay, Thomas Leistner, Jan Metzger, and Oliver C. Schnürer.

Out of these 10 speakers, 7 gave two talks where the first one was more of an introductory nature and the other one was more focused on actual research. These talks covered a broad variety of topics, ranging from special holonomy metrics to various concepts of mass in general relativity and the numerical and analytic study of black hole space-times.

Moreover, three of the speakers gave minicourses where each of them had a total length of 180 min. One minicourse was more of a physical nature and was held by Gary Gibbons about the theory of black holes. The other two lecture courses were more of a mathematical nature. One course was held by Oliver C. Schnürer about geometric flows and focused in particular on mean curvature flow. The other course held by Helga Baum was about special holonomy and parallel spinors in Lorentzian geometry. In addition, we had two related talks about Cauchy problems for Lorentzian manifolds of special holonomy by Thomas Leistner.

This volume consists of two articles. The first is based on the mathematical lecture course by Oliver C. Schnürer and the second on the mathematical lecture course by Helga Baum extended by results presented in the lectures by Thomas Leistner.

Another volume based on the third lecture course about the theory of black holes is planned. The papers are written for graduate students and researchers with a general background in geometry and in the theory of partial differential equations, who want to get acquainted with these central subjects of modern geometry. We hope this volume will be helpful and inspiring.

Hamburg, Germany
July 2018

Vicente Cortés
Klaus Kröncke
Jan Louis

Contents

Lorentzian Geometry: Holonomy, Spinors, and Cauchy Problems

Helga Baum and Thomas Leistner

Contents

H. Baum (✉)
Humboldt University Berlin, Department of Mathematics, Berlin, Germany
e-mail: baum@mathematik.hu-berlin.de

T. Leistner (✉)
School of Mathematical Sciences, University of Adelaide, Adelaide, SA, Australia
e-mail: thomas.leistner@adelaide.edu.au

V. Cortés et al. (eds.), *Geometric Flows and the Geometry of Space-time*, Tutorials, Schools, and Workshops in the Mathematical Sciences,
https://doi.org/10.1007/978-3-030-01126-0_1

Abstract This review is based on lectures given by the authors during the Summer School *Geometric Flows and the Geometry of Space-Time* at the University of Hamburg, September 19–23, 2016. In the first part we describe the algebraic classification of connected Lorentzian holonomy groups. In particular, we specify the holonomy groups of locally indecomposable Lorentzian spin manifolds with a parallel spinor field. In the second part we explain new methods for the construction of globally hyperbolic Lorentzian manifolds with special holonomy based on the solution of certain Cauchy problems for PDEs that are imposed by the existence of a parallel lightlike vector field or a parallel lightlike spinor field with initial conditions on a spacelike hypersurface. Thereby, we derive a second order evolution equation of Cauchy-Kowalevski type that can be solved in the analytic setting as well as an appropriate first order quasilinear hyperbolic system that yields a solution in the smooth case.

1 Introduction

This review is based on lectures given by the authors during the Summer School *Geometric Flows and the Geometry of Space-Time* at the University of Hamburg, September 19–23, 2016. In these lectures we described at one hand the algebraic classification of connected Lorentzian holonomy groups and explained at the other hand new methods for the construction of Lorentzian manifolds with special holonomy based on the solution of appropriate Cauchy problems with initial conditions on a spacelike hypersurface.

The holonomy group of a semi-Riemannian manifold (M, g) is the group of parallel transports along all curves that are closed at a fixed point $x \in M$. It is a Lie subgroup of the group of all orthogonal transformations of $(T_x M, g_x)$, its connected component is isomorphic to the holonomy group of the universal semi-Riemannian covering of (M, g).

The concept of holonomy was probably first successfully applied in differential geometry by E. Cartan [31–33], who used it to classify symmetric spaces. Since then, it has proved to be a very important concept. In particular, it allows to describe parallel sections in geometric vector bundles over (M, g)—such as tangent, tensor or spinor bundles—as holonomy invariant objects and therefore by purely algebraic tools. Moreover, geometric properties like curvature properties can be read off if the holonomy group is special, i.e., a proper subgroup of $\mathrm{O}(T_x M, g_x)$. One of the important consequences of the holonomy notion is its application to the 'classification' of special geometries that are compatible with Riemannian geometry. For each of these geometries an own branch of differential geometry has developed, for example Kähler geometry (holonomy $\mathrm{U}(m)$), geometry of Calabi-Yau manifolds ($\mathrm{SU}(m)$), hyper-Kähler geometry ($\mathrm{Sp}(k)$), quaternionic Kähler geometry ($\mathrm{Sp}(k) \cdot \mathrm{Sp}(1)$), or the exceptional geometry of G_2-manifolds or of $\mathrm{Spin}(7)$-manifolds. In physics there is much interest in semi-Riemannian manifolds with special holonomy, since they often allow to construct spaces with additional supersymmetries (Killing spinors). The development of holonomy theory has a long history. We refer for details to [22, 25, 26, 51].

The *irreducible* holonomy representations of simply connected semi-Riemannian manifolds were classified by M. Berger in the 1950s [19, 20]. Since any holonomy representation of a *Riemannian* manifold splits into irreducible subrepresentations, Berger's results yield the classification of the connected holonomy groups of Riemannian manifolds. The situation in Lorentzian geometry is more difficult. The only connected *irreducible* Lorentzian holonomy group is the group $SO^0(1, n-1)$. Hence, if a connected Lorentzian holonomy group is a proper subgroup of $SO^0(1, n-1)$, then it acts decomposable or it acts indecomposable but non-irreducible, i.e., it admits an invariant degenerate subspace.

The holonomy groups of 4-dimensional Lorentzian manifolds were classified by physicists working in General Relativity [49, 72, 73]. The general dimension was long time ignored. Due to the development of supergravity and string theory in the last decades physicists as well as mathematicians became more interested in higher dimensional Lorentzian geometry. The search for special supersymmetries required the classification of holonomy groups in higher dimension. In the beginning of the 1990s, L. Berard-Bergery and his students began a systematic study of Lorentzian holonomy groups. They discovered many special features of Lorentzian holonomy. Their groundbreaking paper [18] on the algebraic structure of subgroups $H \subset SO^0(1, n-1)$ acting with a degenerate invariant subspace was the starting point for the classification. The second author [60, 61] completed the classification of the connected Lorentzian holonomy groups by the full description of the structure of such $H \subset SO(1, n-1)$ which can appear as holonomy groups. It remained to show that any of the groups in this holonomy list can be realised by a Lorentzian metric. Many realisations were known before but some cases were still open until A. Galaev [44] finally found a realisation for all of the groups.

In the first part of this review we describe these results in more detail. In Sect. 2 we first recall some basic notions of Lorentzian geometry in order to clarify the conventions. For all fundamental differential geometric concepts such as Levi-Civita connection, Lie derivative, etc. we refer to [68]. In Sect. 3 we give a short introduction to holonomy theory of semi-Riemannian manifolds and recall the classification of connected holonomy groups of Riemannian manifolds. Afterwards we explain the classification of connected holonomy groups of Lorentzian manifolds. Special holonomy groups always appear if the manifold is spin and admits a non-trivial parallel spinor field. For this reason we consider in Sect. 4 the relation between holonomy groups and parallel spinor fields. In particular, we discuss the properties of the Ricci curvature of Lorentzian spin manifolds with a parallel spinor field and describe the indecomposable Lorentzian holonomy groups which allow parallel spinors.

In the second part of the review we explain new approaches to construct globally hyperbolic Lorentzian manifolds with special holonomy by solving appropriate Cauchy problems with initial conditions along a spacelike hypersurface based on recent results in [16, 65] and [62], see also [17] for related results. We focus on the case of Lorentzian manifolds which admit non-trivial lightlike parallel vector fields or non-trivial lightlike parallel spinor fields. In both cases the holonomy representation is of special form, it admits an invariant degenerate subspace. At first, in Sect. 5 we derive the necessary constraint conditions, which lightlike parallel vector and spinor fields impose on spacelike hypersurfaces. In the vector field case, the local geometry of Riemannian manifolds satisfying these constraint conditions is completely described. In the spinor field case, the constraint conditions can be expressed as the existence of an so-called imaginary W-Killing spinor of a special algebraic type, where W is the Weingarten operator of the spacelike hypersurface. As an application of the solutions of the Cauchy problem described in Sect. 7 we obtain a local classification of Riemannian manifolds with imaginary W-Killing spinors of this algebraic type (Sect. 8). It is natural to ask whether the constraint conditions for Riemannian manifolds (Σ, h) described in Sect. 5 are not only necessary but also sufficient for (Σ, h) being a Cauchy hypersurface in a Lorentzian manifold with a lightlike parallel vector or spinor field. By studying certain Cauchy problems for PDEs that are induced by the existence of lightlike parallel vector and spinor fields, we show in Sect. 7 that this is indeed the case. Since the methods for the existence of a solution are in part analogous to the approach for the vacuum Einstein equation, we give in Sect. 6 a short review of the approaches for the Einstein equation. After deriving the constraint equations we first describe the vacuum Einstein equation as a second order evolution equation for a family of Riemannian metrics that is of Cauchy-Kowalevski form, that can be solved in the real-analytic setting. Afterwards we explain the method of hyperbolic reduction which allows to consider the vacuum Einstein equation as symmetric hyperbolic system and solve it in the smooth setting. In Sect. 7 we derive in a similar way an evolution equation of Cauchy-Kowalevski type for a parallel lightlike vector field in the analytic setting as well as an appropriate symmetric hyperbolic system which can be solved in the smooth case. Finally we show, that in both cases the solution admits a parallel lightlike spinor field if, in addition, the contraint conditions for parallel spinors on the initial hypersurface are satisfied.

2 Basic Notions

Let (M^n, g) be an n-dimensional manifold[1] with a metric g of signature (p, q), where p denotes the number of -1 and q the number of $+1$ in the normal form of the metric g. We call (M, g) *Riemannian manifold* if $p = 0$, *Lorentzian manifold* if $p = 1 < n$ and *pseudo-Riemannian manifold* if $1 \leq p < n$. If we do not want to specify the signature we use the term *semi-Riemannian manifold.*

Contrary to the Riemannian case, not every manifold admits a Lorentzian metric. There is a topological obstruction (see [68, Chapter 5, Proposition 37] for a proof):

[1] We assume all manifolds to be smooth, connected and without boundary.

Theorem 1 *Let M be a manifold of dimension $n \geq 2$. Then there exists a Lorentzian metric on M if and only if M is non-compact or M is compact with vanishing Euler characteristic.*

Now, let (M, g) be a Lorentzian manifold.

Definition 1 A tangent vector $v \in T_xM$ is called

- timelike, if $g_x(v, v) < 0$,
- spacelike, if $g_x(v, v) > 0$ or $v = 0$,
- lightlike, if $g_x(v, v) = 0$ and $v \neq 0$,
- causal, if v is timelike or lightlike.

Correspondingly, a vector field X is called timelike, spacelike, etc., if $X(x)$ is timelike, spacelike, etc., for all $x \in M$. A a smooth curve $\gamma : I \to M$ is called timelike, spacelike, etc., if all its tangent vectors $\gamma'(t)$ are timelike, spacelike, etc., for all $t \in I$.

Definition 2 Let (M, g) be a Lorentzian manifold. A vector field ξ on M is called *time-orientation* if $g(\xi, \xi) = -1$. If there exists a time-orientation ξ on (M, g), (M, g) is called time-orientable.

A time-oriented Lorentzian manifold is also called spacetime. A time-orientation ξ on a Lorentzian manifold (M, g) singles out one of the two time-cones $\tau^{\pm}(x)$ in any point $x \in M$ in a smooth way, where $\tau^{+}(x)$ and $\tau^{-}(x)$ denote the connected components of $\{v \in T_xM \mid g_x(v, v) < 0\}$. A causal vector field X on M is called *future-directed*, if $g(X, \xi) < 0$, i.e. $X(x)$ and $\xi(x)$ belong to the same time-cone.

In the following we will denote by ∇^g the Levi-Civita connection of (M, g), i.e. the unique metric and torsion free covariant derivative on (M, g). Our convention for the curvature tensor $R^g \in \Gamma(\Lambda^2T^*M \otimes End(TM))$, and $R^g \in \Gamma(\Lambda^2T^*M \otimes \Lambda^2T^*M)$ is the following:

$$R^g(X, Y)Z := \nabla^g_X \nabla^g_Y Z - \nabla^g_Y \nabla^g_X Z - \nabla^g_{[X,Y]} Z,$$
$$R^g(X, Y, Z, W) := g(R^g(X, Y)Z, W).$$

The curvature tensor satisfies the first and second Bianchi-identities,

$$R^g(X, Y)Z + R^g(Y, Z)X + R^g(Z, X)Y = 0,$$
$$\nabla^g_X R^g(Y, Z, U, V) + \nabla^g_Y R^g(Z, X, U, V) + \nabla^g_Z R^g(X, Y, U, V) = 0.$$

Then the Ricci tensor Ric^g and the scalar curvature $scal^g$ of (M, g) are given by

$$Ric^g(X, Y) := \mathrm{tr}_g\, R^g(X, \cdot, \cdot, Y), \qquad scal^g := \mathrm{tr}_g Ric^g.$$

The second Bianchi identity for R^g implies

$$d\, scal^g = 2\mathrm{div}^g(Ric^g), \tag{1}$$

where $\mathrm{div}^g(B) = \mathrm{tr}^g_{(1,2)} \nabla^g B$ denotes the divergence of a bilinear form tensor field B.

For Lorentzian manifolds many classical Theorems of Riemannian geometry, like the Hopf-Rinow Theorem, do not longer hold. For example, there are compact Lorentzian manifolds that are not geodesically complete. An important class of Lorentzian manifolds are the globally hyperbolic spacetimes.

Definition 3 A time-oriented Lorentzian manifold (M, g) is called *globally hyperbolic*, if there exists a Cauchy surface S in M, i.e., a subset $S \subset M$ which is intersected exactly once by any inextendible timelike curve.

Each Cauchy surface $S \subset M$ is an embedded topological hypersurface of M and the topological splitting theorem of R. Geroch [48] states, that M is homeomorphic to $\mathbb{R} \times S$ for any Cauchy surface S. Moreover, there is an important *smooth* splitting theorem for globally hyperbolic manifolds, proven in [21], see also [66, Theorem 3.78].

Theorem 2 *A spacetime (M, g) is globally hyperbolic if and only if it admits a (smooth) spacelike Cauchy hypersurface. In the globally hyperbolic case, for each spacelike Cauchy hypersurface $\Sigma \subset M$, (M, g) is isometric to a Lorentzian manifold of the form*

$$(\mathbb{R} \times \Sigma \, , \, \hat{g} := -\lambda^2 dt^2 + h_t),$$

where $\lambda : \mathbb{R} \times \Sigma \to \mathbb{R}^+$ is a smooth function, called laps function, $(h_t)_{t \in \mathbb{R}}$ is a smooth family of Riemannian metrics on Σ and $\Sigma_t := \{t\} \times \Sigma$ are spacelike Cauchy hypersurfaces for any $t \in \mathbb{R}$.

In some sense, global hyperbolicity is a local property (see for example [6, Lemma A.5.6]).

Lemma 1 *Let (M, g) be a time oriented Lorentzian manifold and $\Sigma \subset M$ a spacelike hypersurface. Then each point on Σ has an open neighbourhood U in M such that $\Sigma \cap U$ is a Cauchy hypersurface in U and hence U is globally hyperbolic.*

Example 1 (Warped Products) Let $I \subset \mathbb{R}$ be an open interval, $f : I \to \mathbb{R}^+$ a smooth function and (Σ, h) a Riemannian manifold. We consider the warped product

$$I \times_f \Sigma := (I \times \Sigma, -dt^2 + f(t)^2 h).$$

Then $I \times_f \Sigma$ is globally hyperbolic if and only if (Σ, h) is complete (for a proof see for example [6, Lemma A.5.14]).

Example 2 (deSitter and Anti-deSitter Spacetime) The Minkowski space is obviously globally hyperbolic. The deSitter spacetime S^n_1, i.e. the simply connected geodesically complete spacetime of constant sectional curvature 1, is globally hyperbolic as well, since it can be described in the form of Example 1 with $I = \mathbb{R}$,

$\Sigma = S^{n-1}$ the Riemannian sphere of radius 1, and $f(t) = \cosh(t)$. Contrary to this, the Anti-deSitter spacetime $\widetilde{H}_1^n$, i.e. the simply connected geodesically complete spacetime of constant sectional curvature -1, is not globally hyperbolic (see [68, Chapter 14, Example 41]).

Globally hyperbolic manifolds have important analytical properties. For example, let E be a vector bundle over a manifold M, and $P : \Gamma(E) \to \Gamma(E)$ a normally hyperbolic operator, i.e., an operator which in local coordinates x^μ on M and a trivialisation of E can be written as

$$P = g^{\mu\nu} \frac{\partial^2}{\partial x^\mu \partial x^\nu} + A^\mu \frac{\partial}{\partial x^\mu} + B,$$

where $g = g_{\mu\nu} dx^\mu dx^\nu$ is a Lorentzian metric, A^μ and B are matrix-valued coefficients depending smoothly on the coordinates x^μ, and where we use Einstein's summation convention. For such operators the *Cauchy problem* is well-posed:

Theorem 3 *If (M, g) is a globally hyperbolic Lorentzian manifold and $P : \Gamma(E) \to \Gamma(E)$ a normally hyperbolic operator, then the Cauchy problem for P is well-posed: if $\Sigma \subset M$ is a spacelike Cauchy hypersurface with the future-directed timelike unit normal field T, then the initial value problem*

$$P\varphi = \psi, \qquad \varphi|_\Sigma = \varphi_0, \qquad \nabla^P_T \varphi|_\Sigma = \varphi_1,$$

has a unique smooth solution $\varphi \in \Gamma(E)$ for given sections $\varphi_0, \varphi_1 \in \Gamma_0(E|_\Sigma)$ and $\psi \in \Gamma_0(E)$. Here ∇^P denotes the P-compatible connection on E and $\Gamma_0(E)$ the smooth sections in E with compact support.

For a proof and further details see the books [6, 40].

3 Lorentzian Holonomy Groups

In this section we will explain the classification of the connected holonomy groups of Lorentzian manifolds. In Sect. 3.1 we first give a short introduction to holonomy theory and recall the classification of the connected holonomy groups of Riemannian manifolds. The proofs of the basic Theorems stated in this subsection can be found in [12, 51, 71]. Then, in Sect. 3.2, we describe the classification of connected Lorentzian holonomy groups.

3.1 Basics on Holonomy Groups

Let (M, g) be a semi-Riemannian manifold of signature (p, q). If $\gamma : [a, b] \to M$ is a piecewise smooth curve connecting two points x and y of M, then for any tangent vector $v \in T_xM$ there is a uniquely determined parallel vector field X_v along γ with initial value v:

$$\frac{\nabla^g X_v}{dt}(t) = 0 \quad \forall\, t \in [a,b], \qquad X_v(a) = v.$$

Since the Levi-Civita connection is compatible with the metric, the parallel transport

$$\begin{aligned} \mathscr{P}^g_\gamma : T_xM &\longrightarrow T_yM \\ v &\longmapsto X_v(b) \end{aligned}$$

defined by X_v is a linear isometry between (T_xM, g_x) and (T_yM, g_y). In particular, if γ is closed, i.e. a loop at x, $\mathscr{P}^g_\gamma$ is an orthogonal linear map on (T_xM, g_x). The *holonomy group of* (M, g) *with respect to* $x \in M$ is the Lie group

$$\mathrm{Hol}_x(M, g) := \{\, \mathscr{P}^g_\gamma : T_xM \to T_xM \mid \gamma \text{ is a loop at } x \,\} \subset \mathrm{O}(T_xM, g_x).$$

Exercise 1 Calculate the holonomy groups for the flat Euclidean space $\mathbb{R}^2$ and for the round sphere $S^2 \subset \mathbb{R}^3$.
Hint for S^2: Consider the following loops at the north pole: go from the north pole along a great circle to the equator, then go a piece along the equator and finally go back to the north pole along a great circle.

If we restrict ourself to null homotopic curves, we obtain the *reduced holonomy group of* (M, g) *with respect to* $x \in M$:

$$\mathrm{Hol}^0_x(M, g) := \{\, \mathscr{P}^g_\gamma : T_xM \to T_xM \mid \gamma \text{ is a null homotopic loop at } x \,\} \subset \mathrm{Hol}_x(M, g).$$

$\mathrm{Hol}^0_x(M, g)$ is the connected component of the identity in the Lie group $\mathrm{Hol}_x(M, g)$. Indeed, contracting a loop γ to one of its points x, gives a curve in $\mathrm{Hol}^0_x(M, g)$ that joins $\mathscr{P}^g_\gamma$ with the identity. Hence, the holonomy group of a simply connected manifold is connected.

The holonomy groups of two different points are conjugated: If σ is a smooth curve connecting x with y, then

$$\mathrm{Hol}_y(M, g) = \mathscr{P}^g_\sigma \circ \mathrm{Hol}_x(M, g) \circ \mathscr{P}^g_{\sigma^{-1}}.$$

Therefore, we often omit the reference point and consider the holonomy groups of (M, g) as class of conjugated subgroups of the (pseudo-)orthogonal group $\mathrm{O}(p, q)$. This requires fixing an orthonormal basis in (T_xM, g_x), changing the basis however does not change this conjugacy class.

If $\pi : (\widetilde{M}, \widetilde{g}) \to (M, g)$ is the universal semi-Riemannian covering, then

$$\mathrm{Hol}^0_{\widetilde{x}}(\widetilde{M}, \widetilde{g}) = \mathrm{Hol}_{\widetilde{x}}(\widetilde{M}, \widetilde{g}) \simeq \mathrm{Hol}^0_{\pi(\widetilde{x})}(M, g).$$

For a semi-Riemannian product $(M, g) = (M_1, g_1) \times (M_2, g_2)$ and $(x_1, x_2) \in M_1 \times M_2$, the holonomy group is the product of its factors

$$\mathrm{Hol}_{(x_1,x_2)}(M, g) = \mathrm{Hol}_{x_1}(M_1, g_1) \times \mathrm{Hol}_{x_2}(M_2, g_2).$$

Exercise 2 Prove the last two statements.

Next, let us describe three general results that relate the holonomy group of (M, g) to geometric properties of (M, g). The first one is the relation to the curvature, described by the *Ambrose Singer Holonomy Theorem*. Due to the symmetry properties of the curvature tensor, for all $x \in M$ and $v, w \in T_xM$ the endomorphism $R^g_x(v, w) : T_xM \to T_xM$ is skew-symmetric with respect to g_x, hence an element of the Lie algebra $\mathfrak{so}(T_xM, g_x)$ of $\mathrm{O}(T_xM, g_x)$. The Lie algebra of the holonomy group is generated by the curvature operators of (M, g), more precisely:

Theorem 4 (Ambrose Singer Holonomy Theorem) *The Lie algebra of the holonomy group* $\mathrm{Hol}_x(M, g)$ *is given by*

$$\mathfrak{hol}_x(M, g) = \mathrm{span}\left\{ (\mathcal{P}^g_\gamma)^{-1} \circ R^g_y\big(v, w\big) \circ \mathcal{P}^g_\gamma \;\middle|\; \begin{array}{l} v, w \in T_yM\,, \\ \gamma \textit{ is a curve from } x \textit{ to } y \end{array} \right\}.$$

This theorem has many important consequences. For example, it tells us that the curvature endomorphisms R^g_x at a point $x \in M$ give a lower bound for the holonomy algebra. On the other hand the holonomy algebra restricts the curvature. In particular, if (M, g) is a locally symmetric space, i.e. $\nabla^g R^g = 0$, then $(\mathcal{P}^g_\gamma)^{-1} \circ R^{\nabla^g}_y\big(\mathcal{P}^g_\gamma(v), \mathcal{P}^g_\gamma(w)\big) \circ \mathcal{P}^g_\gamma = R^g_x(v, w)$, hence

$$\mathfrak{hol}_x(M, g) = \mathrm{span}\{R^g_x(v, w) \mid v, w \in T_xM\}.$$

Finally, the first Bianchi identity is inherited by the operators that span the holonomy algebra and hence poses strong algebraic conditions on the holonomy algebra that are used to derive classification results we will describe below.

Exercise 3 Use the Ambrose-Singer Theorem to calculate the holonomy algebra of the sphere S^n and the hyperbolic space H^n.

If the manifold (M, g) is real analytic, parallel transport is not longer needed to describe the holonomy algebra. It is enough to look at the curvature and at all of its derivatives in one point x.

Theorem 5 *Let* (M, g) *be a real analytic manifold. Then the holonomy algebra of* (M, g) *is spanned by all skew-symmetric endomorphisms*

$$\big(\nabla^g_{v_1} \ldots \nabla^g_{v_k} R^g\big)_x(v, w) : T_xM \to T_xM,$$

where $v, w, v_1, \ldots, v_k \in T_xM$ *and* $0 \le k < \infty$.

A second important property of holonomy groups is stated in the following *holonomy principle*, which relates parallel tensor fields on M to fixed elements under the action of the holonomy group of one point.

Theorem 6 (Holonomy Principle) *Let $\mathcal{T}$ be a tensor bundle on (M, g) and let ∇^g be the covariant derivative on $\mathcal{T}$ induced by the Levi-Civita connection. Then there is a vector space isomorphism between parallel tensor fields $\psi \in \Gamma(\mathcal{T})$ and holonomy invariant tensors $v \in \mathcal{T}_x$ in one point $x \in M$, i.e,*

$$\{\psi \in \Gamma(\mathcal{T}) \mid \nabla^g \psi = 0\} \simeq \{v \in \mathcal{T}_x \mid Hol_x(M, g)v = v\}$$

Proof If $\psi \in \Gamma(\mathcal{T})$ is a tensor field with $\nabla^g \psi = 0$, then $\mathrm{Hol}_x(M, g)\, \psi(x) = \psi(x)$, where $\mathrm{Hol}_x(M, g)$ acts in the canonical way on the tensors $\mathcal{T}_x$. Contrary, if $v \in \mathcal{T}_x$ is a tensor with $\mathrm{Hol}_x(M, g)\, v = v$, then there is an uniquely determined tensor field $\psi \in \Gamma(\mathcal{T})$ with $\nabla^g \psi = 0$ and $\psi(x) = v$. The tensor field ψ is given by parallel transport of v, i.e., $\psi(y) := \mathcal{P}_\gamma^{\nabla^g}(v)$, where $y \in M$ and γ is a curve from x to y. By the holonomy invariance, $\psi(y)$ does not depend on the chosen curve γ. Moreover, ψ is parallel on M, in particular parallel along any smooth curve in M. Hence standard ODE-arguments show, that ψ is smooth. □

Due to this property many intersting special geometric structures can be described by the properties of the holonomy group.

Example 3 Let (M, g) be a Lorentzian manifold. There exists a lightlike (resp. timelike) parallel vector field V on (M, g) if and only if there is a lightlike (resp. timelike) vector $v \in T_x M$ such that $\mathrm{Hol}_x(M, g)v = v$.

Example 4 (M^{2m}, g) is a Kähler manifold if and only if $\mathrm{Hol}_x(M, g) \subset U(m)$.

To see this, remember that $\mathrm{U}(m)$ is embedded in the group $\mathrm{SO}(2m)$ by

$$A + iB \in \mathrm{U}(m) \mapsto \begin{pmatrix} A & -B \\ B & A \end{pmatrix} \in \mathrm{SO}(2m).$$

Using this embedding, $\mathrm{U}(m)$ is described as the stabilizer of the standard almost complex structure $J_0 = \begin{pmatrix} 0 & -\mathrm{I}_m \\ \mathrm{I}_m & 0 \end{pmatrix} \in \mathrm{End}(\mathbb{R}^{2m})$, where $\mathrm{SO}(2m)$ acts by conjugation on $\mathrm{End}(\mathbb{R}^{2m})$. Therefore, the holonomy group satisfies $\mathrm{Hol}_x(M, g) \subset \mathrm{U}(m) \subset \mathrm{SO}(2m)$ if and only if $\mathrm{Hol}_x(M, g)J_0 = J_0$. By the holonomy principle this is equivalent to the existence of a parallel, orthogonal, almost complex structure J on (M, g), given by the parallel transport of J_0, i.e. (M, g) is a Kähler manifold.

Exercise 4 Prove that a semi-Riemannian manifold (M, g) of signature (p, q) is orientable if and only if $\mathrm{Hol}_x(M, g) \subset \mathrm{SO}(p, q)$.

The third important relation between holonomy and geometry are local spitting properties. For a subspace $E \subset T_x M$ we denote by

$$E^\perp = \{v \in T_x M \mid g_x(v, E) = 0\} \subset T_x M$$

its orthogonal complement. If E is holonomy invariant, i.e. $\mathrm{Hol}_x(M,g)E \subset E$, then $E^{\perp}$ is holonomy invariant as well. If E is in addition non-degenerate, then $E^{\perp}$ is non-degenerate as well and T_xM is the direct sum of these holonomy invariant subspaces:

$$T_xM = E \oplus E^{\perp}.$$

Theorem 7 (Local and Global Splitting Theorem) *Let $E \subset T_xM^n$ be a proper, non-degenerate, holonomy invariant subspace of T_xM of dimension k. Then (M,g) is locally a metric product, i.e. for each point $y \in M$ there exists an open neighborhood $U(y)$ such that $(U(y), g|_{U(y)})$ is isometric to a product of two semi-Riemannian manifolds of dimension k and $n-k$ respectively*

$$(U(y), g|_{U(y)}) \overset{\text{isometric}}{\simeq} (U_1, g_1) \times (U_2, g_2).$$

Moreover, $\mathrm{Hol}^0_x(M,g)$ is isomorphic to the product of two groups $H_1 \times H_2$, where $H_1 \subset \mathrm{O}(E)$ and $H_2 \subset \mathrm{O}(E^{\perp})$.

If in addition, (M,g) is simply connected and geodesicalls complete, then (M,g) is globally isometric to a product of two semi-Riemannian manifolds

$$(M,g) \overset{\text{isometric}}{\simeq} (M_1, g_1) \times (M_2, g_2)$$

with

$$\mathrm{Hol}_x(M,g) \simeq \mathrm{Hol}_{x_1}(M_1, g_1) \times \mathrm{Hol}_{x_2}(M_2, g_2).$$

The local decomposition of (M,g) follows from the Frobenius Theorem. If $E \subset T_xM$ is a non-degenerate, holonomy invariant subspace, then

$$\mathscr{E} : y \in M \longrightarrow \mathscr{E}_y := \mathscr{P}^g_\sigma(E) \subset T_yM,$$

where σ is a piecewise smooth curve from x to y, is an involutive distribution on M, the *holonomy distribution defined by E*. The maximal connected integral manifolds of $\mathscr{E}$ are totally geodesic submanifolds of M, which are geodesically complete if (M,g) is so. The manifolds (U_1, g_1) and (U_2, g_2) in Theorem 7 can be chosen as small open neighborhood of y in the integral manifold $M_1(y)$ of the holonomy distribution $\mathscr{E}$ defined by E and the integral manifold $M_2(y)$ of the holonomy distribution $\mathscr{E}^{\perp}$ defined by $E^{\perp}$, respectively, with thc mctrics induced by g. If (M,g) is simply connected and geodesically complete, (M,g) is even globally isometric to the product of the two integral manifolds $(M_1(x), g_1)$ and $(M_2(x), g_2)$.

The holonomy group $\mathrm{Hol}_x(M,g)$ acts as group of orthogonal linear mappings on the tangent space (T_xM, g_x). This representation is called the *holonomy representation of* (M,g), we denote it in the following by ρ. The holonomy representation $\rho : \mathrm{Hol}_x(M,g) \to \mathrm{O}(T_xM, g_x)$ is called *irreducible* if there is no proper holonomy invariant subspace $E \subset T_xM$, and *indecomposable*, if there is no proper *non-degenerate* holonomy invariant subspace $E \subset T_xM$. To be short, we say that the holonomy group or its Lie algebra acts *irreducibly (indecomposably)*, if the holonomy representation has this property. If (M, g) is a Riemannian manifold, irreducible is the same as indecomposable. In the pseudo-Riemannian case there are indecomposable holonomy representations which admit degenerate holonomy invariant subspaces, i.e., which are not irreducible. This causes the problems in the classification of the holonomy groups of pseudo-Riemannian manifolds.

In view of Theorem 7 we call a semi-Riemannian manifold (M, g) *irreducible*, if the holonomy representation of $\mathrm{Hol}^0(M, g)$ is irreducible, and *(locally) indecomposable*, if it is indecomposable.

Now, we decompose the tangent space T_xM into a direct sum of non-degenerate, orthogonal and holonomy invariant subspaces

$$T_xM = E_0 \oplus E_1 \oplus \ldots \oplus E_r,$$

where $\mathrm{Hol}_x(M, g)$ acts indecomposable on $E_1, \ldots, E_r$ and E_0 is a maximal non-degenerate subspace (possibly 0-dimensional), on which the holonomy group acts trivial. Applying Theorem 7 to this decomposition we obtain the *Decomposition Theorem of de Rham and Wu* [35, 77].

Theorem 8 (De Rham-Wu Decomposition Theorem) *Let (M, g) be a simply connected, geodesically complete semi-Riemannian manifold. Then (M, g) is isometric to a product of simply connected, geodesically complete semi-Riemannian manifolds*

$$(M, g) \simeq (M_0, g_0) \times (M_1, g_1) \times \ldots \times (M_r, g_r),$$

where (M_0, g_0) is a (possibly 0-dimensional) (pseudo-)Euclidian space and the factors $(M_1, g_1), \ldots, (M_r, g_r)$ are indecomposable and non-flat.

Theorem 8 reduces the classification of connected holonomy groups of geodesically complete semi-Riemannian manifolds to the study of indecomposable holonomy representations. This classification is widely open, only the case of *irreducible* holonomy representations is completely solved for every signature.

First of all, let us mention that the holonomy group of a symmetric space is given by its isotropy representation, a result that goes back to Cartan.

Theorem 9 *Let (M, g) be a symmetric space, and let $G(M) \subset \mathrm{Isom}(M, g)$ be its transvection group. Furthermore, let $\lambda : H(M) \longrightarrow \mathrm{GL}(T_{x_0}M)$ be the isotropy representation of the stabiliser $H(M) = G(M)_{x_0}$ of a point $x_0 \in M$. Then,*

$$\lambda(H(M)) = \mathrm{Hol}_{x_0}(M, g).$$

In particular, the holonomy group $\mathrm{Hol}_{x_0}(M, g)$ *is isomorphic to the stabilizer* $H(M)$ *and, using this isomorphism, the holonomy representation* ρ *is given by the isotropy representation* λ.

Therefore, the holonomy groups of symmetric spaces can be read off from the classification lists of symmetric spaces, which describe the pair $(G(M), H(M))$ and the isotropy representation λ. For *irreducible* symmetric spaces these lists can be found in [22, Chapter 10], in [50] and in [20]. In order to classify the *irreducible* holonomy representations, the classification of the non-symmetric case remains. This was done by M. Berger in 1955 [19]. He proved that there is only a short list of groups which can appear as holonomy groups of *irreducible* non-locally symmetric simply connected semi-Riemannian manifolds. This list is now called the *Berger list*. The Berger list of *Riemannian* manifolds is well-known. There appear only six special holonomy groups and due to the holonomy principle (Theorem 6) each of these groups is related to a special, rich and interesting geometry, described by the corresponding parallel geometric object. For more details on the corresponding geometries, see [22] or [52].

Theorem 10 (Riemannian Berger List) *Let* (M^n, g) *be an n-dimensional, irreducible Riemannian manifold. Then the connected holonomy group* $\mathrm{Hol}^0(M, g)$ *is up to conjugation in* $\mathrm{O}(n)$ *one of the following groups with its standard representation,*

n	*Holonomy group*	*Special geometry*
n	$\mathrm{SO}(n)$	—
$2m \geq 4$	$\mathrm{U}(m)$	*Kähler manifold*
$2m \geq 4$	$\mathrm{SU}(m)$	*Ricci-flat Kähler manifold*
$4m \geq 8$	$\mathrm{Sp}(m)$	*Hyperkähler manifold*
$4m \geq 8$	$\mathrm{Sp}(m) \cdot \mathrm{Sp}(1)$	*quaternionic Kähler manifold*
7	G_2	G_2-manifold
8	$\mathrm{Spin}(7)$	$\mathrm{Spin}(7)$-manifold

or $(\mathrm{Hol}^0(M, g), \rho)$ *is the isotropy representation of a simply connected irreducible Riemannian symmetric space.*

M. Berger also classified the connected holonomy groups of irreducible *pseudo-Riemannian* manifolds. In the Lorentzian case, this list does not contain a proper subgroup of $\mathrm{SO}^0(1, n-1)$. This reflects a special algebraic fact concerning irreducibly acting connected subgroups of the Lorentzian group $\mathrm{O}(1, n-1)$ (see for example [36]).

Theorem 11 *If* $H \subset \mathrm{O}(1, n-1)$ *is a connected Lie subgroup acting irreducibly on* $\mathbb{R}^{1,n-1}$, *then* $H = \mathrm{SO}^0(1, n-1)$.

3.2 Holonomy Groups of Lorentzian Manifolds

In this section we will describe the algebraic classification of the connected holonomy groups of Lorentzian manifolds.

In dimension 4 there are 14 types of Lorentzian holonomy groups which were first listed by J.F. Schell [72] and R. Shaw [73], see also [22, Chapter 10]. We will not recall this list here, instead, in the following we will consider arbitrary dimension and take a more systematic approach (for other surveys of the following results, see [45, 46]).

Due to Theorems 8 and 11 the decomposition theorem for Lorentzian manifolds can be formulated as follows:

Theorem 12 *Let (N, h) be a simply connected, geodesically complete Lorentzian manifold. Then (N, h) is isometric to the product*

$$(N, h) \simeq (M, g) \times (M_1, g_1) \times \ldots \times (M_r, g_r),$$

where (M_i, g_i) are either flat or irreducible Riemannian manifolds and (M, g) is either

1. $(\mathbb{R}, -dt^2)$,
2. *an n-dimensional irreducible Lorentzian manifold with holonomy group* $\mathrm{Hol}^0(M, g) \simeq \mathrm{SO}^0(1, n-1)$, *or*
3. *an indecomposable, non-irreducible Lorentzian manifold.*

Since the holonomy groups of the Riemannian factors are known, it remains to classify the indecomposable, non-irreducible Lorentzian holonomy representations.

Corollary 1 *Let (M^n, g) be an indecomposable, non-irreducible Lorentzian manifold. Then the holonomy representation $\rho : \mathrm{Hol}^0_x(M, g) \to \mathrm{O}(T_xM, g_x)$ admits a degenerate invariant subspace $W \subset T_xM$. The intersection $L := W \cap W^\perp \subset T_xM$ is a lightlike line, which is also invariant under the* full *holonomy group* $\mathrm{Hol}_x(M, g)$. *In particular, the holonomy group $\mathrm{Hol}_x(M, g)$ lies in the stabilizer $\mathrm{O}(T_xM, g_x)_L$ of L in $\mathrm{O}(T_xM, g_x)$:*

$$\mathrm{Hol}_x(M, g) \subset \mathrm{O}(T_xM, g_x)_L.$$

For the proof that the invariance of L under $\mathrm{Hol}^0_x(M, g)$ implies its invariance under the full $\mathrm{Hol}_x(M, g)$ and for interesting examples with $\mathrm{Hol}^0_x(M, g) \neq \mathrm{Hol}_x(M, g)$, see [15]. Geometrically, this means that (M^n, g) admits an lightlike parallel line bundle $\mathbb{V} \subset TM$, defined by

$$\mathbb{V}_y := \mathbb{R}\mathscr{P}^g_\gamma(L),$$

where γ is a smooth curve from x to y. Moreover, $\mathbb{V} \subset \mathbb{V}^{\perp}$. Since $\mathbb{V}$ and $\mathbb{V}^{\perp}$ are integrable distributions, (M, g) is foliated into totally geodesic lightlike hypersurfaces, each itself foliated by lightlike pregeodesics.

Let us describe the stabilizer $O(T_xM, g_x)_L$ in detail. For that we fix a Witt basis $(s_-, s_1, \ldots, s_{n-2}, s_+)$ in (T_xM, g_x) such that $s_- \in L$ and

$$\Big(g_x(s_\alpha, s_\beta)\Big) = \begin{pmatrix} 0 & 0 & 1 \\ 0 & I_{n-2} & 0 \\ 1 & 0 & 0 \end{pmatrix}, \quad \text{where } \alpha, \beta \in \{-, 1, \ldots, n-2, +\}, \tag{2}$$

identify (T_xM, g_x) with the Minkowski space and and write the elements of $O(T_xM, g_x)$ as matrices with respect to this basis. The stabilizer of the lightlike line $L = \mathbb{R}s_- \subset \mathbb{R}^{1,n-1}$ is a semidirect product and given by the matrices

$$\begin{aligned} O(1, n-1)_L &= \big(\mathbb{R}^* \times O(n-2)\big) \ltimes \mathbb{R}^{n-2} \\ &= \left\{ \begin{pmatrix} a^{-1} & x^t & -\frac{1}{2}a\|x\|^2 \\ 0 & A & -aAx \\ 0 & 0 & a \end{pmatrix} \;\middle|\; \begin{array}{l} a \in \mathbb{R}^* \\ x \in \mathbb{R}^{n-2} \\ A \in O(n-2) \end{array} \right\}. \end{aligned}$$

The Lie algebra of $O(1, n-1)_L$ is

$$\begin{aligned} \mathfrak{so}(1, n-1)_L &= \big(\mathbb{R} \oplus \mathfrak{so}(n-2)\big) \ltimes \mathbb{R}^{n-2} \\ &= \left\{ \begin{pmatrix} r & y^t & 0 \\ 0 & X & -y \\ 0 & 0 & -r \end{pmatrix} \;\middle|\; \begin{array}{l} r \in \mathbb{R} \\ y \in \mathbb{R}^{n-2} \\ X \in \mathfrak{so}(n-2) \end{array} \right\}. \end{aligned}$$

If we describe a matrix in the Lie algebra $\mathfrak{so}(1, n-1)_L$ by (r, X, y) (in the obvious way), the commutator is given by

$$[(r, X, y), (s, Y, z)] = \big(0, [X, Y], (X + r\,\mathrm{Id})z - (Y + s\,\mathrm{Id})y\big).$$

In particular, $\mathbb{R}$, $\mathbb{R}^{n-2}$ and $\mathfrak{so}(n-2)$ are subalgebras of $\mathfrak{so}(1, n-1)_L$. Let $\mathfrak{h}$ be a subalgebra of $\mathfrak{so}(1, n-1)_L$. We call the subalgebra

$$\mathfrak{g} := \mathrm{proj}_{\mathfrak{so}(n-2)}(\mathfrak{h}) \subset \mathfrak{so}(n-2)$$

the *orthogonal part of* $\mathfrak{h}$. It is reductive, i.e., its Levi decomposition is given by $\mathfrak{g} = \mathfrak{z}(\mathfrak{g}) \oplus [\mathfrak{g}, \mathfrak{g}]$, where $\mathfrak{z}(\mathfrak{g})$ is the center of $\mathfrak{g}$ and the commutator $[\mathfrak{g}, \mathfrak{g}]$ is semi-simple.

The first step in the classification of indecomposable, non-irreducible holonomy representations is a result due to L. Berard-Bergery and A. Ikemakhen [18], who described the possible algebraic types of indecomposable, non-irreducible subalgebras $\mathfrak{h}$ of the stabilizer $\mathfrak{so}(1,n-1)_L = \big(\mathbb{R}\oplus\mathfrak{so}(n-2)\big)\ltimes\mathbb{R}^{n-2}$. A geometric proof of this result was later given by A. Galaev in [43].

Theorem 13 *Let $L \subset \mathbb{R}^{1,n-1}$ be a lightlike line in the Minkowski space, let*

$$\mathfrak{h} \subset \mathfrak{so}(1, n-1)_L = (\mathbb{R} \oplus \mathfrak{so}(n-2)) \ltimes \mathbb{R}^{n-2}$$

be an indecomposable subalgebra and let $\mathfrak{g} = \mathrm{proj}_{\mathfrak{so}(n-2)}(\mathfrak{h}) = \mathfrak{z}(\mathfrak{g}) \oplus [\mathfrak{g}, \mathfrak{g}]$ be its orthogonal part. Then $\mathfrak{h}$ is of one of the following four types:

1. $\mathfrak{h}^1(\mathfrak{g}) := (\mathbb{R} \oplus \mathfrak{g}) \ltimes \mathbb{R}^{n-2}$.
2. $\mathfrak{h}^2(\mathfrak{g}) := \mathfrak{g} \ltimes \mathbb{R}^{n-2}$.
3. $\mathfrak{h}^3(\mathfrak{g}, \varphi) := \big\{\big(\varphi(X), X+Y, z\big) \mid X \in \mathfrak{z}(\mathfrak{g}),\ Y \in [\mathfrak{g},\mathfrak{g}],\ z \in \mathbb{R}^{n-2}\big\}$,
   where $\varphi : \mathfrak{z}(\mathfrak{g}) \to \mathbb{R}$ is a linear and surjective map.
4. $\mathfrak{h}^4(\mathfrak{g}, \psi) := \big\{\big(0, X+Y, \psi(X)+z\big) \mid X \in \mathfrak{z}(\mathfrak{g}),\ Y \in [\mathfrak{g},\mathfrak{g}],\ z \in \mathbb{R}^{k}\big\}$,
   where $\mathbb{R}^{n-2} = \mathbb{R}^m \oplus \mathbb{R}^k\ ,\ 0 < m < n-3$,
   $\mathfrak{g} \subset \mathfrak{so}(\mathbb{R}^k)$,
   $\psi : \mathfrak{z}(\mathfrak{g}) \to \mathbb{R}^m$ *is linear and surjective.*

In the following we will refer to theses cases as the Lie algebras $\mathfrak{h}$ of type 1 to type 4. The types 1 and 2 are called *uncoupled types*, the types 3 and 4 *coupled types*, since the $\mathfrak{so}(n-2)$-part is coupled by φ and ψ with the $\mathbb{R}$- and the $\mathbb{R}^{n-2}$-part, respectively. If the holonomy algebra $\mathfrak{h} = \mathfrak{hol}_x(M,g)$ is of type type 2 or 4 (in this case $\mathrm{proj}_{\mathbb{R}}(\mathfrak{h}) = 0$), the universal covering of (M,g) admits a parallel lightlike vector field V. If the holonomy algebra is of type 1 or 3 (i.e., with $\mathrm{proj}_{\mathbb{R}}(\mathfrak{h}) \neq 0$), the universal covering admits a recurrent lightlike vector field V, i.e. $\nabla^{\widetilde{g}} V = \eta \otimes V$ with a 1-form η such that $d\eta \neq 0$. The orthogonal part $\mathfrak{g}$ of $\mathfrak{hol}(M,g)$ has the following geometric meaning. The Levi-Civita connection ∇^g induces a covariant derivative $\nabla^{\mathbb{S}}$ on the vector bundle $\mathbb{S} := \mathbb{V}^{\perp}/\mathbb{V}$ over M by

$$\nabla^{\mathbb{S}}_X[Y] := [\nabla^g_X Y], \qquad \text{where } Y \in \Gamma(\mathbb{V}^{\perp}) \text{ and } X \in \mathfrak{X}(M).$$

$\mathbb{S}$ is the so-called *screen bundle* and we will come back to it in Sect. 8. It is not difficult to show, that the holonomy algebra of $(\mathbb{S}, \nabla^{\mathbb{S}})$ coincides with $\mathfrak{g}$.

Moreover, the different types of holonomy algebras translate into special curvature properties of the lightlike hypersurface of M, defined by the involutive distribution $\mathcal{V}^{\perp}$. For details we refer to [23].

For a classification of the holonomy algebras $\mathfrak{hol}(M,g)$ one has to give a description of its orthogonal parts $\mathfrak{g}$. This was done by the second author who obtained the following result [59–61].

Theorem 14 *Let (M^n, g) be an indecomposable, non-irreducible Lorentzian manifold. Then the orthogonal part $\mathfrak{g} = \mathrm{proj}_{\mathfrak{so}(n-2)}(\mathfrak{hol}(M,g))$ of the holonomy algebra is the holonomy algebra of a Riemannian manifold (with its holonomy representation).*

The proof of this theorem is based on the observation, that the orthogonal part of a Lorentzian holonomy algebra has a special algebraic property. It is a so-called *weak Berger algebra*—a notion, which was introduced and studied by the second author in [59] (see also [44, 61]). We will explain this notion here shortly.

Let $\mathfrak{g} \subset \mathfrak{gl}(V)$ be a subalgebra of the linear maps of a finite dimensional real or complex vector space V with scalar product $\langle \cdot, \cdot \rangle$. Then we consider the following spaces

$$\mathscr{K}(\mathfrak{g}) := \{R \in \Lambda^2(V^*) \otimes \mathfrak{g} \mid R(x, y)z + R(y, z)x + R(z, x)y = 0\},$$
$$\mathscr{B}(\mathfrak{g}) := \{B \in V^* \otimes \mathfrak{g} \mid \langle B(x)y, z\rangle + \langle B(y)z, x\rangle + \langle B(z)x, y\rangle = 0\}.$$

The space $\mathscr{K}(\mathfrak{g})$ is called *the space of algebraic curvature tensors of* $\mathfrak{g}$. This name is motivated by the fact, that the condition which defines $\mathscr{K}(\mathfrak{g})$ is just the Bianchi identity for the curvature tensor R^{∇}_x of a torsion free covariant derivative ∇. The space $\mathscr{B}(\mathfrak{g})$ is called *space of weak algebraic curvature tensors of* $\mathfrak{g}$. A Lie algebra $\mathfrak{g} \subset \mathfrak{gl}(V)$ is called *Berger algebra* if there are enough algebraic curvature tensors to generate $\mathfrak{g}$, i.e., if

$$\mathfrak{g} = \operatorname{span}\{R(x, y) \mid x, y \in V,\ R \in \mathscr{K}(\mathfrak{g})\}.$$

An orthogonal Lie algebra $\mathfrak{g} \subset \mathfrak{so}(V, \langle \cdot, \cdot \rangle)$ is called *weak Berger algebra* if there are enough weak algebraic curvature tensors to generate $\mathfrak{g}$, i.e., if

$$\mathfrak{g} = \operatorname{span}\{B(x) \mid x \in V,\ B \in \mathscr{B}(\mathfrak{g})\}.$$

Obviously, every orthogonal Berger algebra $\mathfrak{g} \subset \mathfrak{so}(V, \langle \cdot, \cdot \rangle)$ is a weak Berger algebra. If $(V, \langle \cdot, \cdot \rangle)$ is an Euclidian space and $\mathfrak{g} \subset \mathfrak{so}(V)$ a weak Berger algebra, then V decomposes into orthogonal $\mathfrak{g}$-invariant subspaces

$$V = V_0 \oplus V_1 \oplus \ldots \oplus V_s, \tag{3}$$

where $\mathfrak{g}$ acts trivial on V_0 (possibly 0-dimensional) and irreducible on V_j, $j = 1, \ldots, s$. Moreover, $\mathfrak{g}$ is the direct sum of ideals

$$\mathfrak{g} = \mathfrak{g}_1 \oplus \ldots \oplus \mathfrak{g}_s,$$

where $\mathfrak{g}_j$ acts irreducible on V_j and trivial on V_i if $i \neq j$. $\mathfrak{g}_j \subset \mathfrak{so}(V_j)$ is a weak Berger algebra and $\mathscr{B}(\mathfrak{g}) = \mathscr{B}(\mathfrak{g}_1) \oplus \ldots \oplus \mathscr{B}(\mathfrak{g}_s)$.

Now, let (M, g) be an indecomposable, non-irreducible Lorentzian manifold with holonomy group $\mathrm{Hol}_x(M, g)$. From the Ambrose-Singer Theorem 4 and the first Bianchi identity it follows that the holonomy algebra $\mathfrak{hol}_x(M, g)$ is a Berger algebra.[2] Moreover, the orthogonal part $\mathfrak{g}$ of $\mathfrak{hol}_x(M, g)$ is a weak Berger algebra of an Euclidian space, hence it decomposes into a direct sum of irreducibly acting weak Berger algebras. Using representation theory of semi-simple Lie algebras, the second author classified all irreducible weak Berger algebras and consequently showed that any irreducible weak Berger algebra on an Euclidian space is the holonomy algebra of an irreducible Riemannian manifold. This implies Theorem 14.

It remains the question whether there are further restrictions for the holonomy algebra of an indecomposable, non-irreducible Lorentzian manifold. A. Galaev proved that this is not the case. In fact, *any* of the algebras in the list of Theorem 13, where $\mathfrak{g}$ is the holonomy algebra of a Riemannian manifold, can be realised as holonomy algebra of a Lorentzian manifold. To show this, A. Galaev constructed a real analytic Lorentzian metric g on $\mathbb{R}^n$, such that $\mathfrak{hol}_0(\mathbb{R}^n, g)$ is of the form $\mathfrak{h}^1(\mathfrak{g})$, $\mathfrak{h}^2(\mathfrak{g})$, $\mathfrak{h}^3(\mathfrak{g}, \varphi)$ or $\mathfrak{h}^4(\mathfrak{g}, \psi)$, as described in Theorem 13, with $\mathfrak{g}$ a holonomy algebra of a Riemannian manifold. To describe these metrics, let us fix some notations. As we know, $\mathbb{R}^{n-2}$ has a decomposition into orthogonal subspaces

$$\mathbb{R}^{n-2} = \mathbb{R}^{n_0} \times \mathbb{R}^{n_1} \times \ldots \times \mathbb{R}^{n_s}, \tag{4}$$

where $\mathfrak{g}$ acts trivial on $\mathbb{R}^{n_0}$ and irreducible on $\mathbb{R}^{n_1}, \ldots, \mathbb{R}^{n_s}$. Now, let $(e_1, \ldots, e_{n-2})$ be an orthonormal basis of $\mathbb{R}^{n-2}$ adapted to the decomposition (4). We choose weak algebraic curvature endomorphisms $Q_\alpha \in \mathscr{B}(\mathfrak{g})$, $\alpha = 1, \ldots, N$, which generate $\mathscr{B}(\mathfrak{g})$.

In case of $\mathfrak{z}(\mathfrak{g}) \neq 0$, we extend the surjective linear maps $\varphi : \mathfrak{z}(\mathfrak{g}) \to \mathbb{R}$ and $\psi : \mathfrak{z}(\mathfrak{g}) \to \mathbb{R}^m \subset \mathbb{R}^{n_0}$ to $\mathfrak{g}$ by setting $\varphi|_{[\mathfrak{g},\mathfrak{g}]} = 0$, $\psi|_{[\mathfrak{g},\mathfrak{g}]} = 0$ and define the numbers:

$$\varphi_{\alpha i} := \frac{1}{(\alpha-1)!}\varphi(Q_\alpha(e_i))$$

$$\psi_{\alpha i j} := \frac{1}{(\alpha-1)!}\Big\langle \psi(Q_\alpha(e_i)), e_j \Big\rangle_{\mathbb{R}^{n-2}},$$

where $\alpha = 1, \ldots, N,\ i = n_0 + 1, \ldots, n-2,\ j = 1, \ldots, m$. Using the description of the holonomy algebra of an real analytic manifold in Theorem 5, A. Galaev proved in [44] the following theorem (see also [45]).

Theorem 15 *Let* $\mathfrak{h} \subset \mathfrak{so}(1, n-1)$ *be one of the Lie algebras* $\mathfrak{h}^1(\mathfrak{g})$, $\mathfrak{h}^2(\mathfrak{g})$, $\mathfrak{h}^3(\mathfrak{g}, \varphi)$, $\mathfrak{h}^4(\mathfrak{g}, \psi)$ *in the list of Theorem 13, where* $\mathfrak{g} = \mathrm{proj}_{\mathfrak{so}(n-2)}\mathfrak{h} \subset \mathfrak{so}(n-2)$ *is the holonomy algebra of a Riemannian manifold. We consider the following metric* g

[2] In fact, the holonomy algebra of every torsionfree connection is a Berger algebra.

on $\mathbb{R}^n$ with coordinates $(v, u, x_1, \dots, x_{n-2})$:

$$g = 2dvdu + 2\sum_{i=1}^{n-2} A_i dx_i du + f\,du^2 + \sum_{i=1}^{n-2} dx_i^2,$$

where the functions A_i are given by

$$A_i(u, x_1, \dots, x_{n-2}) := \sum_{\alpha=1}^{N}\sum_{k,l=1}^{n-2} \tfrac{1}{3(\alpha-1)!}\langle Q_\alpha(e_k)e_l + Q_\alpha(e_l)e_k, e_i\rangle_{\mathbb{R}^{n-2}} x_k x_l u^\alpha,$$

and the function $f(v, u, x_1, \dots, x_{n-2})$ is defined in the following list, corresponding to the type of $\mathfrak{h}$:

$\mathfrak{h}$	f
Type 1: $\mathfrak{h}^1(\mathfrak{g}) = (\mathbb{R} \oplus \mathfrak{g}) \ltimes \mathbb{R}^{n-2}$	$v^2 + \sum_{i=1}^{n_0} x_i^2$
Type 2: $\mathfrak{h}^2(\mathfrak{g}) = \mathfrak{g} \ltimes \mathbb{R}^{n-2}$	$\sum_{i=1}^{n_0} x_i^2$
Type 3: $\mathfrak{h}^3(\mathfrak{g}, \varphi)$	$2v \sum_{\alpha=1}^{N} \sum_{i=n_0+1}^{n-2} \varphi_{\alpha i} x_i u^{\alpha-1} + \sum_{k=1}^{n_0} x_k^2$
Type 4: $\mathfrak{h}^4(\mathfrak{g}, \psi)$	$2 \sum_{\alpha=1}^{N} \sum_{i=n_0+1}^{n-2} \sum_{j=1}^{m} \psi_{\alpha ij} x_i x_j u^{\alpha-1} + \sum_{k=m+1}^{n_0} x_k^2$

Then, $\mathfrak{h}$ is the holonomy algebra of $(\mathbb{R}^n, g)$ with respect to the point $0 \in \mathbb{R}^n$.

Finally, we describe further examples of Lorentzian manifolds with special holonomy.

Example 5 Let (M^n, g) be a simply connected indecomposable Lorentzian symmetric space. The classification of these spaces was obtained by Cahen and Wallach in [29].

a) If (M^n, g) is irreducible, then (M^n, g) is isometric to a simply connected space form of constant non-zero sectional curvature and $\mathrm{Hol}(M^n, g) = SO^0(1, n-1)$.
b) If (M, g) is non-irreducible, then (M^n, g) is isometric to a Cahen-Wallach space $CW^n(\underline{\lambda})$ of dimension $n \geq 3$, which is the $\mathbb{R}^n$ with the metric

$$g_{\underline{\lambda}} := 2dvdu + \sum_{i=1}^{n-2} \lambda_i x_i^2 du^2 + \sum_{i=1}^{n-2} dx_i^2,$$

where $\underline{\lambda} = (\lambda_1, \ldots, \lambda_{n-2})$ is a tuple of non-zero real parameters. In this case, the holonomy group is abelian, $\mathrm{Hol}(M^n, g) = \mathbb{R}^{n-2}$.

Example 6 Let (N, h) be an $(n-2)$-dimensional Riemannian manifold, and let $f \in C^\infty(\mathbb{R}^2 \times N)$ be a smooth function such that the Hessian of $f(0, 0, \cdot) \in C^\infty(N)$ is non-degenerate in $p_0 \in N$. Then the holonomy group of the Lorentzian manifold $(M, g^{f,h})$

$$M := \mathbb{R}^2 \times N \ , \ \ g^{f,h} := 2dvdu + f\,du^2 + h,$$

where v, u denote the coordinates of $\mathbb{R}^2$, is given by

$$\mathrm{Hol}_{(0,0,p_0)}(M, g^{f,h}) = \begin{cases} \mathrm{Hol}_{p_0}(N, h) \ltimes \mathbb{R}^{n-2} & if \ \ \frac{\partial f}{\partial v} = 0, \\ (\mathbb{R}^+ \times \mathrm{Hol}_{p_0}(N, h)) \ltimes \mathbb{R}^{n-2} & if \ \ \frac{\partial^2 f}{\partial v^2} \neq 0. \end{cases}$$

This can be proved by direct calculation of the group of parallel transports (cf. for example [12], chap. 5).

Example 7 Let (B, h_B) be a Riemannian manifold and $\pi : M \to B$ an S^1-principal bundle over B. Choose a connection form A on M, a smooth real function f on M and a closed nowhere vanishing 1-form η on B. Then

$$g := 2i\,A \circ \pi^*\eta + f(\pi^*\eta)^2 + \pi^* h_B$$

is a Lorentzian metric. The fundamental vertical vector field V, induced by i in the Lie algebra if S^1, is lightlike and recurrent,

$$\nabla^g_X V = -V(f)\eta(d\pi(X))\,V.$$

One can choose the topological type of the S^1-bundle, A, f and η in such a way, that (M, g) is indecomposable and non-irreducible with holonomy group of type 1 and 2 (cf. K. Lärz in [57]). Lärz used this construction to derive geodesically complete examples, compact examples and totally twisted examples (i.e. examples without topological splitting of a 1-dimensional factor as in the previous examples).

An important source for Lorentzian manifolds with special holonomy group are spin manifolds admitting parallel spinor fields. In the next section we will explain this in detail.

4 Lorentzian Spin Geometry: Curvature and Holonomy

In this section we will give a short introduction to Lorentzian spin manifolds and discuss consequences of the existence of parallel spinor fields on Lorentzian manifolds for curvature properties and the shape of the holonomy group.

4.1 Spin Structures and Spinor Fields

We start with an introduction of spin structures and spinor fields on semi-Riemannian manifolds.

Definition 4 Let (M, g) be a semi-Riemannian manifold of signature (p, q). Then (M, g) is called *space- and time-orientable*, if the bundle of all adapted[3] orthonormal frames of (M, g) can be reduced to the connected subgroup $SO^0(p, q) \subset O(p, q)$. Such a reduction will be denoted by $\mathscr{F}(M, g)$ and called space- and time-orientation.

For a Riemannian manifold space- and time-orientability is the same as orientability.

Exercise 5 Prove, that a Lorentzian manifold (M, g) is space- and time-orientable if and only if M is orientable and (M, g) admits a time-orientation ξ. In this case, the fibre of $\mathscr{F}(M, g)$ in $x \in M$ is given by

$$\mathscr{F}(M, g)_x := \left\{ s_x := (s_1, \ldots, s_n) \;\middle|\; \begin{array}{l} s_x \text{ positively oriented orthonormal basis in } (T_xM, g_x), \\ s_1 \text{ timelike and future directed, } s_2, \ldots, s_n \text{ spacelike.} \end{array} \right\}.$$

Now let $\langle \cdot, \cdot \rangle_{p,q}$ be the bilinear form on $\mathbb{R}^n$, $n = p + q$, given by

$$\langle v, w \rangle_{p,q} := -v_1 w_1 - \ldots - v_p w_p + v_{p+1} w_{p+1} + \ldots + v_n w_n,$$

where $v = (v_1, \ldots, v_n)$, $w = (w_1, \ldots, w_n) \in \mathbb{R}^n$. We denote by $\mathrm{Cliff}_{p,q}$ the Clifford algebra of the space $(\mathbb{R}^n, -\langle \cdot, \cdot \rangle_{p,q})$. $\mathrm{Cliff}_{p,q}$ is a real associative algebra with $\mathbf{1}$ generated by $\mathbb{R}^n$ with the relations

$$v \cdot w + w \cdot v = -2 \langle v, w \rangle_{p,q} \mathbf{1}.$$

As vector space it is isomorphic to the exterior algebra $\Lambda^* \mathbb{R}^n$. There is an important group $\mathrm{Spin}(p, q) \subset \mathrm{Cliff}_{p,q}$, the *spin group*, defined by

$$\mathrm{Spin}(p, q) := \{ v_1 \cdot \ldots \cdot v_{2k} \mid \langle v_j, v_j \rangle_{p,q} = \pm 1, \; j = 1, \ldots, 2k, \; k \in \mathbb{N} \}.$$

$\mathrm{Spin}(p, q)$ is a Lie group, the connected component of the identity, $\mathrm{Spin}^0(p, q)$, is given by the products $v_1 \cdot \ldots \cdot v_{2k}$ with even number of timelike vectors v_j. In the positive definite case, the Spin group $\mathrm{Spin}(n) := \mathrm{Spin}(0, n)$ is connected. The Lie algebra $\mathfrak{spin}(p, q)$ of $Spin(p, q)$ can be described as subspace of $\mathrm{Cliff}_{p,q}$

$$\mathfrak{spin}(p, q) = \mathrm{span}\{ e_i \cdot e_j \mid 1 \leq i < j \leq n \} \simeq \Lambda^2 \mathbb{R}^n \subset \mathrm{Cliff}_{p,q},$$

[3] We call an orthonormal basis $(s_1, \ldots, s_n)$ of (T_xM, g_x) *adapted*, if $g_x(s_j, s_j) = -1$ for $1 \leq j \leq p$ and $g_x(s_j, s_j) = 1$ for $p + 1 \leq j \leq n$.

where $(e_1, \ldots, e_n)$ denotes the canonical basis of $\mathbb{R}^n$, with the Lie bracket

$$[X, Y] := X \cdot Y - Y \cdot X, \qquad X, Y \in \mathfrak{spin}(p, q).$$

There is a twofold covering $\lambda : \mathrm{Spin}(p, q) \to \mathrm{SO}(p, q)$, defined by

$$\lambda(v_1 \cdot \ldots \cdot v_{2k}) := S_{v_1} \circ \ldots \circ S_{v_{2k}},$$

where the map S_v for $v \in \mathbb{R}^n$, $\langle v, v\rangle_{p,q} = \pm 1$, denotes the reflection across the hypersurface $v^\perp \subset \mathbb{R}^n$,

$$S_v(w) = w - 2\frac{\langle v, w\rangle_{p,q}}{\langle v, v\rangle_{p,q}} v, \qquad w \in \mathbb{R}^n.$$

For index $p = 0$, the twofold covering λ is universal iff $n \geq 3$, for index $p = 1$, λ is the universal iff $n \geq 4$.

The complexified Clifford algebra $\mathrm{Cliff}^{\mathbb{C}}_{p,q}$ has exactly one equivalence class of irreducible representations if $n = p + q$ is even. If n is odd, there are exactly two such equivalence classes. Restricting this irreducible representation to $\mathrm{Spin}(p, q) \subset \mathrm{Cliff}^{\mathbb{C}}_{p,q}$ one obtains the fundamental representation of the spin group, the *spin representation*[4]

$$\kappa : \mathrm{Spin}(p, q) \longrightarrow \mathrm{GL}(\Delta_{p,q}).$$

The representation space $\Delta_{p,q}$ is a complex vector space of dimension $2^{[n/2]}$. $\Delta_{p,q}$ is an irreducible $\mathrm{Spin}(p, q)$-module, if n is odd, and splits into two irreducible submodules $\Delta_{p,q} = \Delta^+_{p,q} \oplus \Delta^-_{p,q}$, if n is even (for proofs see [58, Ch.I.5]). Since $\mathbb{R}^n \subset \mathrm{Cliff}_{p,q}$, we can multiply vectors from $\mathbb{R}^n$ with elements of $\Delta_{p,q}$. This multiplication, called *Clifford multiplication*, is $\mathrm{Spin}(p, q)$-equivariant. Moreover, there is a $\mathrm{Spin}^0(p, q)$-invariant hermitian form $\langle\cdot,\cdot\rangle_{\Delta_{p,q}}$ on $\Delta_{p,q}$, which is positive definite, if $p = 0, n$, and indefinite of split signature, if $1 \leq p < n$. An explicit realisation of the spin representation, the Clifford multiplication and the form $\langle\cdot,\cdot\rangle_{\Delta_{p,q}}$ can be found for example in [8, 11] or [41].

Definition 5 Let (M, g) be a time- and space-oriented semi-Riemannian manifold. A *spin structure* of (M, g) is a pair $(\widetilde{\mathscr{F}}(M, g), \Lambda)$ of a principal fibre bundle $\widetilde{\mathscr{F}}(M, g)$ over M with structure group $\mathrm{Spin}^0(p, q)$ and a smooth and surjective map $\Lambda : \widetilde{\mathscr{F}}(M, g) \to \mathscr{F}(M, g)$ which respects the bundle projections as well as the group actions, i.e.

$$\pi_{\widetilde{\mathscr{F}}} = \pi_{\mathscr{F}} \circ \Lambda,$$
$$\Lambda(q \cdot a) = \Lambda(q) \cdot \lambda(a), \qquad \text{for all } q \in \widetilde{\mathscr{F}}(M, g) \text{ and } a \in \mathrm{Spin}^0(p, q).$$

[4] If $n = p + q$ is odd, the two possible restrictions are equivalent as $Spin(p, q)$-representations.

Definition 6 A semi-Riemannian manifold is called *spin manifold*, if it is space- and time-oriented and if it admits a spin structure.

Not every space- and time-oriented semi-Riemannian manifold admits a spin structure. There is a topological obstruction that ensure the existence of spin structures, namely $w_2(M) = 0$, where w_2 is the second Stiefel-Whitney class, see [8, 55]. If $w_2(M) = 0$, the number of non-equivalent spin structures coincides with the number of elements in $H^1(M, \mathbb{Z}_2)$. On spin manifolds, besides of tensor bundles, there exists a special complex vector bundle, the *spinor bundle* and special differential operators acting on its smooth sections, the *spinor fields*, which are of interests in Mathematics as well as in Physics, like the Dirac operator or the twistor operator. We shortly recall the basic properties of the differential calculus for spinor fields.

Let (M, g) be a semi-Riemannian spin manifold of dimension n and signature (p, q) with spin structure $(\widetilde{\mathscr{F}}(M, g), f)$. Using the spin representation $\Delta_{p,q}$, we can define *the spinor bundle*, i.e. the associated complex vector bundle

$$S := \widetilde{\mathscr{F}}(M, g) \times_{\mathrm{Spin}^0(p,q)} \Delta_{p,q}.$$

The Clifford multiplication and the scalar product $\langle\cdot,\cdot\rangle_{\Delta_{p,q}}$ on $\Delta_{p,q}$ are $\mathrm{Spin}^0(p,q)$-invariant. Hence, we can extend these fibre-wise algebraic objects to the bundles over M and obtain:

1. the *Clifford multiplication* at the bundle level

$$\begin{aligned} \mu : TM \otimes S &\longrightarrow S \\ X \otimes \varphi &\longmapsto X \cdot \varphi \end{aligned} \qquad \text{and} \qquad \begin{aligned} \mu^* : T^*M \otimes S &\longrightarrow S\,, \\ \omega \otimes \varphi &\longmapsto \omega \cdot \varphi := \omega^\sharp \cdot \varphi \end{aligned}$$

2. the non-degenerate hermitian *bundle metric* $\langle\cdot\,\cdot\rangle_S$ on S,
3. the *spinor derivative* $\nabla^S : \Gamma(S) \to \Gamma(T^*M \otimes S)$, i.e. the covariant derivative on S induced by the Levi-Civita connection ∇^g, locally given by

$$\nabla^S_X \varphi = X(\varphi) + \frac{1}{2} \sum_{k<l} g(\nabla^g_X s_k, s_l)\, \sigma^k \cdot \sigma^l \cdot \varphi, \tag{5}$$

where $(s_1, \dots, s_n)$ is a local orthonormal frame and $(\sigma^1, \dots, \sigma^n)$ the dual frame.

Thereby, the following rules are satisfied for all vector fields X, Y on M and all spinor fields $\varphi, \psi \in \Gamma(S)$:

$$(X \cdot Y + Y \cdot X) \cdot \varphi = -2\, g(X, Y)\, \varphi, \tag{6}$$

$$\langle X \cdot \varphi, \psi \rangle_S = (-1)^{p+1} \langle \varphi, X \cdot \psi \rangle_S, \tag{7}$$

$$\nabla^S (X \cdot \varphi) = \nabla^g (X) \cdot \varphi + X \cdot \nabla^S \varphi, \tag{8}$$

$$X(\langle \varphi, \psi \rangle_S) = \langle \nabla^S_X \varphi, \psi \rangle_S + \langle \varphi, \nabla^S_X \psi \rangle_S. \tag{9}$$

The Dirac operator $D : \Gamma(S) \to \Gamma(S)$ on a semi-Riemannian spin manifold is given by $D := \mu^* \circ \nabla^S$, locally,

$$D\varphi = \sum_{j=1}^{n} \sigma^j \cdot \nabla^S_{s_j}\varphi, \quad \varphi \in \Gamma(S).$$

A spinor field $\varphi \in \Gamma(S)$ is called *harmonic*, if $D\varphi = 0$, and *parallel*, if $\nabla^S\varphi = 0$.

To each spinor field φ belongs a special vector field, the Dirac current of φ.

Definition 7 Let $\varphi \in \Gamma(S)$ be a spinor field on a semi-Riemannian spin manifold (M, g) of index p. Then the vector field V_φ on M, given by

$$g(V_\varphi, X) := i^{p+1}\langle\varphi, X \cdot \varphi\rangle_S \quad \text{for all } X \in \mathfrak{X}(M),$$

is called *Dirac current of* φ.

The Dirac current V_φ is a real vector field, since by relation (7) we have

$$\overline{i^{p+1}\langle\varphi, X \cdot \varphi\rangle_S} = (-i)^{p+1}\langle X \cdot \varphi, \varphi\rangle_S = i^{p+1}\langle\varphi, X \cdot \varphi\rangle_S.$$

Proposition 1 *If $\varphi \in \Gamma(S)$ is harmonic, the Dirac current V_φ is divergence free. If $\varphi \in \Gamma(S)$ is parallel, the Dirac current V_φ is parallel as well.*

Proof We use the rules (7)–(9) for the spinor calculus and obtain for the derivative of the Dirac current V_φ:

$$\begin{aligned}
g(\nabla^g_Y V_\varphi, X) &= Y\big(g(V_\varphi, X)\big) - g(V_\varphi, \nabla^g_Y X)\\
&= i^{p+1}Y\big(\langle\varphi, X \cdot \varphi\rangle_S\big) - i^{p+1}\langle\varphi, \nabla^g_Y X \cdot \varphi\rangle_S\\
&\overset{(9)}{=} i^{p+1}\langle\nabla^S_Y\varphi, X \cdot \varphi\rangle_S + i^{p+1}\langle\varphi, \nabla^S_Y(X \cdot \varphi)\rangle_S - i^{p+1}\langle\varphi, \nabla^g_Y X \cdot \varphi\rangle_S\\
&\overset{(8)}{=} i^{p+1}\langle\nabla^S_Y\varphi, X \cdot \varphi\rangle_S + i^{p+1}\langle\varphi, X \cdot \nabla^S_Y\varphi\rangle_S\\
&\overset{(7)}{=} (-i)^{p+1}\langle X \cdot \nabla^S_Y\varphi, \varphi\rangle_S + i^{p+1}\langle\varphi, X \cdot \nabla^S_Y\varphi\rangle_S\\
&= 2\,Re\big((-i)^{p+1}\langle X \cdot \nabla^S_Y\varphi, \varphi\rangle_S\big).
\end{aligned}$$

It immediately follows that the Dirac current of a parallel spinor field is parallel as well. For the divergence of V_φ we obtain

$$\begin{aligned}
div^g(V_\varphi) &= \sum_{k=1}^{n} \varepsilon_k g(\nabla^g_{s_k} V_\varphi, s_k) = 2Re\big((-i)^{p+1}\big\langle \sum_{k=1}^{n} \varepsilon_k s_k \cdot \nabla^S_{s_k}\varphi, \varphi\big\rangle_S\big)\\
&= 2Re\big((-i)^{p+1}\langle D\varphi, \varphi\rangle_S\big),
\end{aligned}$$

where $(s_1, \dots, s_n)$ denotes a local orthonormal frame and $\varepsilon_k := g(s_k, s_k) = \pm 1$. Hence, the Dirac current of a harmonic spinor field is divergence free. □

4.2 *Curvature and Holonomy of Lorentzian Manifolds with Parallel Spinors*

Now, let us focus on *Lorentzian* spin manifolds (M, g). Whereas for Riemannian spin manifolds, the bundle metric $\langle \cdot, \cdot \rangle_S$ is positive definite and the Clifford multiplication $X\cdot : S \to S$ is a skew symmetric map, in case of Lorentzian spin manifolds, $\langle \cdot, \cdot \rangle_S$ is *indefinite* of split signature and $X\cdot : S \to S$ is symmetric. Using a time-orientation ξ of (M, g), we obtain a *positive definite* hermitian bundle metric $(\cdot, \cdot)_\xi$ on S by

$$(\varphi, \psi)_\xi := \langle \xi \cdot \varphi, \psi \rangle_S,$$

but the price one has to pay is that the Clifford multiplication does not behaves so nicely, namely,

$$(X \cdot \varphi, \psi)_\xi = -(\varphi, \theta(X) \cdot \psi)_\xi,$$

where $\theta : TM \to TM$ is the reflection across $\xi^\perp$ (cf. [8] or [11]).

The Dirac current of a spinor field on a Lorentzian spin manifold has special properties.

Proposition 2 *The Dirac current of a spinor field φ on a Lorentzian spin manifold satisfies*

1. $\text{zero}(\varphi) = \text{zero}(V_\varphi)$.
2. $g(V_\varphi, V_\varphi) \leq 0$.
3. V_φ *is future directed on* $M \setminus \text{zero}(V_\varphi)$.

Proof Let ξ be a time-orientation of (M, g). Then

$$g(V_\varphi, \xi) = -\langle \xi \cdot \varphi, \varphi \rangle_S = -(\varphi, \varphi)_\xi \leq 0. \tag{10}$$

Since $(\cdot, \cdot)_\xi$ is positive definite, this shows, that $V_\varphi(x) = 0$ implies $\varphi(x) = 0$. On the other hand, by the very definition of V_φ, $\varphi(x) = 0$ implies $V_\varphi(x) = 0$. Hence, the zero sets of V_φ and φ coincide. Equation (10) shows in addition, that V_φ is future-directed on $M \setminus \text{zero}(V_\varphi)$.

We now decompose V_φ in a timelike and a spacelike part:

$$V_\varphi =: -g(V_\varphi, \xi)\xi + Z.$$

Then $Z \in \xi^\perp$ is spacelike and $Z(x) = 0$ implies $g_x(V_\varphi(x), V_\varphi(x)) \leq 0$. It remains to show, that $g(V_\varphi, V_\varphi) \leq 0$ on the open submanifold $\widetilde{M} := M \setminus (\mathrm{zero}(Z) \cup \mathrm{zero}(V_\varphi))$ of M. On $\widetilde{M}$ we consider the spacelike unit vector field

$$N := -\frac{Z}{\sqrt{g(Z, Z)}} \in \xi^\perp.$$

Then on $\widetilde{M}$

$$V_\varphi = \alpha\, \xi - \beta\, N, \tag{11}$$

where $\alpha := -g(V_\varphi, \xi) > 0$ and $\beta := \sqrt{g(Z, Z)} > 0$.

The bundle map $\xi \cdot N \cdot : S_{|\widetilde{M}} \to S_{|\widetilde{M}}$ is an involution, since

$$(\xi \cdot N) \cdot (\xi \cdot N) \cdot = -\underbrace{(\xi \cdot \xi)}_{\mathbf{1}} \cdot \underbrace{(N \cdot N)}_{\mathbf{-1}} \cdot = \mathrm{Id}_{S_{|\widetilde{M}}}.$$

Let $S_{\pm 1}$ be the eigenspaces of $\xi \cdot N \cdot$ to the eigenvalues ± 1. Then

$$S_{|\widetilde{M}} = S_1 \oplus S_{-1}$$

is an orthogonal decomposition with respect to $(\cdot, \cdot)_\xi$ and

$$\varphi \in \Gamma(S_{\pm 1}) \iff \xi \cdot \varphi = \pm N \cdot \varphi. \tag{12}$$

Now, consider the decomposition of $\varphi \in \Gamma(S_{|\widetilde{M}})$ with respect of these eigenspaces, $\varphi = \varphi_1 + \varphi_{-1} \in \Gamma(S_1 \oplus S_{-1})$. Then

$$\alpha = -g(V_\varphi, \xi) \overset{(10)}{=} (\varphi, \varphi)_\xi = \|\varphi_1\|_\xi^2 + \|\varphi_{-1}\|_\xi^2. \tag{13}$$

$$\begin{aligned} \beta &= -g(V_\varphi, N) = \langle N \cdot \varphi, \varphi \rangle_S = (\xi \cdot N \cdot \varphi, \varphi)_\xi = (\varphi_1 - \varphi_{-1}, \varphi_1 + \varphi_{-1})_\xi \\ &= \|\varphi_1\|_\xi^2 - \|\varphi_{-1}\|_\xi^2. \end{aligned} \tag{14}$$

Hence on $\widetilde{M}$,

$$g(V_\varphi, V_\varphi) = -\alpha^2 + \beta^2 = -4\|\varphi_1\|_\xi^2\, \|\varphi_{-1}\|_\xi^2 \leq 0.$$

□

Remark 1 The Dirac current V_ψ of a spinor field $\psi \in \Gamma(S)$ on a *Riemannian* spin manifold satisfies $\mathrm{zero}(\psi) \subset \mathrm{zero}(V_\psi)$. Contrary to the *Lorentzian* case, in general equality does not hold, $V_\psi = 0$ can happen for a non-zero spinor field ψ.

Definition 8 We call a spinor field on a Lorentzian spin manifold *lightlike*, if the Dirac current V_φ is lightlike.

Proposition 3 *Let $\varphi \in \Gamma(S)$ be a nowhere vanishing spinors field on a Lorentzian spin manifold. Then φ is lightlike if and only if $V_\varphi \cdot \varphi = 0$.*

Proof Let $\varphi = \varphi_1 + \varphi_{-1} \in \Gamma(S_1 \oplus S_{-1})$ as in the proof of Proposition 2. We use the decomposition $V_\varphi \overset{(11)}{=} \alpha\xi - \beta N$ on M and obtain

$$\begin{aligned} g(V_\varphi, V_\varphi) = 0 &\iff -\alpha^2 + \beta^2 = 0 \\ &\iff \alpha = \beta > 0 \\ &\overset{(13),(14)}{\iff} \|\varphi_1\|_\xi^2 + \|\varphi_{-1}\|_\xi^2 = \|\varphi_1\|_\xi^2 - \|\varphi_{-1}\|_\xi^2 \\ &\iff \varphi_{-1} = 0 \ \text{ i.e. } \varphi \in \Gamma(S_1) \\ &\overset{(12)}{\iff} \xi \cdot \varphi = N \cdot \varphi \\ &\iff V_\varphi \cdot \varphi = \alpha\,(\xi - N) \cdot \varphi = 0. \end{aligned}$$

□

Applying the holonomy principle (Theorem 6) and the local splitting theorem (Theorem 7), we obtain for the holonomy group $\mathrm{Hol}_x(M, g)$:

Corollary 2 *Let $\varphi \in \Gamma(S)$ be a parallel spinor field, $\varphi \neq 0$. Then the parallel vector field V_φ is either lightlike or timelike with $g(V_\varphi, V_\varphi) = const < 0$. In particular, $\mathrm{Hol}_x(M, g)$ is contained in the stabilizer of $V_\varphi(x)$ in $\mathrm{O}(T_xM, g_x)$,*

$$\mathrm{Hol}_x(M, g) \subset \mathrm{O}(T_xM, g_x)_{V_\varphi(x)} \subset \mathrm{O}(T_xM, g_x).$$

a) If V_φ is timelike, then

$$T_xM = \mathbb{R}\, V_\varphi(x) \oplus V_\varphi(x)^\perp$$

is a holonomy-invariant decomposition into non-degenerate subspaces, hence, (M, g) is locally isometric to a metric product $(\mathbb{R} \times F, -dt^2 + g_F)$, where (F, g_F) is a Riemannian spin manifold with a parallel spinor field.

b) If V_φ is lightlike, then $\mathrm{Hol}_x(M, g)$ lies in the stabilizer of a lightlike vector, hence

$$\mathrm{Hol}_x(M, g) \subset \mathrm{SO}(n-2) \ltimes \mathbb{R}^{n-2} \subset \mathrm{SO}^0(1, n-1).$$

A Riemannian spin manifold with parallel spinors is Ricci-flat (see Remark 2 below). In contrast, for Lorentzian spin manifolds with parallel spinor fields the Ricci tensor may be non-zero. Here is an example where this happens.

Exercise 6 We consider the Cahen-Wallach spaces $CW^n(\underline{\lambda}) = (\mathbb{R}^n, g_{\underline{\lambda}})$, with

$$g_{\underline{\lambda}} = 2dvdu + \sum_{j=1}^{n-2} \lambda_j x_j^2 du^2 + \sum_{j=1}^{n-2} dx_j^2,$$

from Example 5. Prove:

1. $V := \frac{\partial}{\partial v}$ is lightlike and parallel.
2. $Ric^g(X,Y) = -\Big(\sum_{j=1}^{n-2} \lambda_j\Big)\, g(X,V)g(Y,V)$ for all vector fields X and Y.
3. $CW^n(\underline{\lambda})$ is a spin manifold and the dimension of the space of parallel spinors is half of the rank of the spinor bundle, i.e. $2^{[n/2]-1}$.
4. If φ is a parallel spinor field on $CW^n(\underline{\lambda})$, then $V_\varphi = \|\varphi\|_\xi^2\, V$ for an appropriate chosen time-orientation ξ.

In general, for Lorentzian spin manifolds we have:

Proposition 4 *Let $\varphi \in \Gamma(S)$ be a parallel, lightlike spinor field on a Lorentzian spin manifold (M, g). Then there exists a smooth function f on M such that*

$$Ric^g = f\; V_\varphi^\flat \otimes V_\varphi^\flat \qquad \textit{and} \qquad V_\varphi(f) = 0. \tag{15}$$

Moreover, the scalar curvature of (M, g) vanishes, $scal^g = 0$.

Proof Using the local formula (5) for the spinor derivative one can calculate for the curvature of ∇^S

$$R^S(X,Y)\varphi = \Big(\Big[\nabla^S_X, \nabla^S_Y\Big] - \nabla^S_{[X,Y]}\Big)\,\varphi = \tfrac{1}{2} \sum_{k<l} R^g(X,Y,s_k,s_l)\,\sigma^k \cdot \sigma^l \cdot \varphi. \tag{16}$$

The symmetry properties of the curvature tensor R^g yield the following formulas for the Ricci tensor and the scalar curvature:

$$\sum_{j=1}^{n} \sigma^j \cdot R^S(X, s_j)\,\varphi = -\frac{1}{2}\, Ric^g(X) \cdot \varphi, \tag{17}$$

$$\sum_{j=1}^{n} \sigma^j \cdot Ric^g(s_j) \cdot \varphi = scal^g\, \varphi. \tag{18}$$

In particular, if φ is parallel and non-zero, (17) and (18) imply

$$Ric^g(X) \cdot \varphi = 0 \quad \text{for all vector fields } X, \tag{19}$$

$$scal^g = 0. \tag{20}$$

Now, let φ be parallel and *lightlike*. Then, in addition to (19), $V_\varphi \cdot \varphi = 0$. Hence, using (6) we obtain

$$Ric^g(X) \cdot Ric^g(X) \cdot \varphi = -g(Ric^g(X), Ric^g(X))\,\varphi = 0,$$
$$\big(Ric^g(X) \cdot V_\varphi + V_\varphi \cdot Ric^g(X)\big) \cdot \varphi = -2g(Ric^g(X), V_\varphi)\,\varphi = 0.$$

Therefore,

$$0 = g(V_\varphi, V_\varphi) = g(Ric^g(X), Ric^g(X)) = g(V_\varphi, Ric^g(X)).$$

Since g is a Lorentzian metric, V_φ and $Ric^g(X)$ have to be linearly dependent, i.e. there exists a 1-form ω on M such that $Ric^g(X) = \omega(X) \cdot V_\varphi$, or for the (2, 0)-Ricci tensor,

$$Ric^g(X, Y) = \omega(X) \cdot g(V_\varphi, Y). \tag{21}$$

Now, let us consider the timelike vector field $T := \frac{\xi}{g(\xi, V_\varphi)}$, satisfying $g(T, V_\varphi) = 1$, and the function $f := \omega(T)$. Then,

$$\omega(X) = \omega(X)\, g(V_\varphi, T) = Ric^g(X, T) = Ric^g(T, X) = \omega(T)\, g(V_\varphi, X) = f\, g(V_\varphi, X).$$

Hence,

$$Ric^g(X, Y) = f\, g(V_\varphi, X)\, g(V_\varphi, Y) \quad \text{for all vector fields } X, Y,$$

which proves the first part of (15). Now, since the scalar curvature of (M, g) vanishes, (1) implies

$$0 = \mathrm{div}^g(Ric^g) = \mathrm{div}^g(f\ V_\varphi^\flat \otimes V_\varphi^\flat) = V_\varphi(f) + f\,\mathrm{div}^g V_\varphi \cdot V_\varphi^\flat + f\ (\nabla^g_{V_\varphi} V_\varphi)^\flat.$$

Since V_φ is parallel, this shows the second part of (15). □

Remark 2 In the Riemannian case, formula (19) implies, that a Riemannian spin manifold with a non-trivial parallel spinor field is Ricci-flat. Hence, by the local splitting result in Corollary 2a), if a Lorentzian spin manifold admits a parallel spinor field with timelike Dirac current, it is Ricci flat as well.

Next, we adress the question how the holonomy groups of Lorentzian spin manifolds with parallel spinor fields look like.

The relation between parallel spinor fields and holonomy groups is based on the following observation [67, 75]:

Proposition 5 *Let (M, g) be a space- and time-oriented semi-Riemannian manifold of signature (p, q). Then, there exists a spin structure on (M, g) with a non-vanishing parallel spinor field if and only if there exists a homomorphism*

$\ell : \mathrm{Hol}(M, g) \to \mathrm{Spin}^0(p, q)$ *with* $\lambda \circ \ell = \mathrm{Id}_{\mathrm{Hol}(M,g)}$ *and a non-vanishing spinor* $v \in \Delta_{p,q}$, *fixed under* $\mathrm{Hol}(M, g)$, *i.e.* ,

$$\ell(\mathrm{Hol}(M, g))v = v.$$

Moreover, the dimension of the space of parallel spinor fields on (M, g) *and the dimension of the space of fixed spinors under* $\mathrm{Hol}(M, g)$ *coincide.*

It was shown by M.Y. Wang [75] that the connected holonomy group $\mathrm{Hol}^0(M, g)$ of an irreducible Riemannian spin manifold with parallel spinor fields is one of the groups $\mathrm{SU}(n/2)$, $\mathrm{Sp}(n/4)$, G_2 or $\mathrm{Spin}(7)$. Since Riemannian spin manifolds with parallel spinor fields are Ricci-flat and Ricci-flat homogeneous Riemannian spaces are flat (see [1]), one only has to check the criteria of Proposition 5 for the six proper subgroups of $\mathrm{SO}(n)$ in the Riemannian Berger list (Theorem 10 above). Wang also determined the possible full holonomy groups $\mathrm{Hol}(M, g)$ of irreducible Riemannian spin manifolds with parallel spinor fields [76], which are the groups G described in the following algebraic lemma (see also [15]).

Lemma 2 *Let* $G \subset \mathrm{SO}(n)$ *be a Lie group with connected component* G^o *equal to* $\mathrm{SU}(n/2)$, $\mathrm{Sp}(n/4)$, G_2 *(n=7) or* $\mathrm{Spin}(7)$ *(n=8) and with a non-zero fixed spinor in* Δ_n. *Then G is equal to one of the groups in the following table, in which N is the dimension of the space of spinors fixed under G:*

G^0	n	G	N	Conditions
$\mathrm{SU}(m)$	$2m$	$\mathrm{SU}(m)$	2	
		$\mathrm{SU}(m) \rtimes \mathbb{Z}_2$	1	*m divisible by* 4
$\mathrm{Sp}(k)$	$4k$	$\mathrm{Sp}(k)$	$k+1$	
		$\mathrm{Sp}(k) \times \mathbb{Z}_d$	$(k+1)/d$	$d > 1$, *d odd and divides* $k+1$
		$\mathrm{Sp}(k) \cdot \mathbb{Z}_{2d}$	$2\lfloor \frac{k}{2d} \rfloor + 1$	*k even*, $1 < d \leq 2d$
		$\mathrm{Sp}(k) \cdot Q_{4d}$	$\lfloor \frac{k}{2d} \rfloor$ *if* $\frac{k}{2}$ *odd*	*k even*, $1 < d \leq 2d$,
			$\lfloor \frac{k}{2d} \rfloor + 1$ *if* $\frac{k}{2}$ *even*	
		$\mathrm{Sp}(k) \cdot B_{4d}$	*see Ref. [76]*	*k even and conditions in* [76]
		$\mathrm{Sp}(k) \cdot \Gamma$	1	*k even*
Spin_7	8	Spin_7	1	
G_2	7	G_2	1	

Here

1. Q_{4d} *is the double cover of the dihedral group* D_{2d} *of order* $2d$,
2. $\mathrm{Sp}(k) \cdot B_{4d}$ *for* $d = 6, 12, 30$, *and* B_{4d} *is the double cover in* $\mathrm{Sp}(1)$ *of the polyhedral groups* P_{2d} *in* $\mathrm{SO}(3)$, *i.e. the tetrahedral group* P_{12}, *the octahedral group* P_{24}, *and the icosahedral group* P_{60}, *and*
3. Γ *is an infinite subgroup of* $\mathrm{U}(1) \rtimes \mathbb{Z}_2$.

In order to describe the holonomy groups of Lorentzian spin manifolds with parallel spinor fields the following algebraic fact is useful.

Lemma 3 *Let* $H = G \ltimes \mathbb{R}^{n-2} \subset \mathrm{SO}^0(1, n-1)$ *with* $G \subset \mathrm{SO}(n-2)$*. Then the following spaces have the same dimension*

1. *spinors in* Δ_{n-2} *fixed under* G*.*
2. *spinors in* $\Delta_{1,n-1}$ *fixed under* H*.*

Proof If $\ell : H = G \ltimes \mathbb{R}^{n-2} \to \mathrm{Spin}^0(1, n-1)$ is a homomorphism with $\lambda \circ \ell = \mathrm{Id}_H$, the restriction of ℓ to G maps into $\lambda^{-1}(\mathrm{SO}(n-2)) = \mathrm{Spin}(n-2) \subset \mathrm{Spin}^0(1, n-1)$, hence we can define $\ell_0 := \ell|_G$ and obtain $\lambda \circ \ell_0 = \mathrm{Id}_G$. $\mathbb{R}^{n-2} \subset H$ is a connected closed Abelian subgroup. When realizing $\mathrm{Spin}^0(1, n-1)$ in the Clifford algebra $\mathrm{Cliff}_{1,n-1}$, the image of $\mathbb{R}^{n-2}$ under ℓ is given by

$$\ell(\mathbb{R}^{n-2}) = \{1 + e_- \cdot x \mid x \in \mathbb{R}^{n-2}\} \subset \mathrm{Spin}^0(1, n-1) \subset \mathrm{Cliff}_{1,n-1},$$

where $(e_-, e_1, \ldots, e_{n-2}, e_+)$ is a Witt basis of the Minkowski space $\mathbb{R}^{1,n-1}$ as in (2). Then, if $\ell_0 : G \to \mathrm{Spin}(n-2)$ is a homomorphism with $\lambda \circ \ell_0 = \mathrm{Id}_\mathrm{G}$, we define $\ell : H = G \ltimes \mathbb{R}^{n-2} \to \mathrm{Spin}^0(1, n-1)$ by

$$\ell(g \cdot x) := \ell_0(g) \cdot (1 + e_- \cdot x), \qquad g \in G,\ x \in \mathbb{R}^{n-2}.$$

It remains to check the conditions for the fixed spinors. We proceed as in the proof of Proposition 2. Let e_0, e_{n-1} be the orthonormal basis in $\mathrm{span}(e_-, e_+)$ given by $e_0 := \frac{1}{\sqrt{2}}(e_- - e_+)$ and $e_{n-1} := \frac{1}{\sqrt{2}}(e_- + e_+)$. Then $e_0 \cdot e_{n-1} \cdot$ is an involution on the spinor modul $\Delta_{1,n-1}$. Let $\Delta_{1,n-1} = \Delta_+ \oplus \Delta_-$ be the decomposition into its the eigenspaces to the eigenvalues ± 1. Then $\Delta_\pm|_{\mathrm{Spin}(n-2)}$ is isomorphic to the spin representation Δ_{n-2}, $w \in \Delta_\pm$ iff $e_0 \cdot w = \pm e_{n-1} \cdot w$ and $e_{n-1} \cdot \Delta_\pm = \Delta_\mp$. Hence for $v := v_+ + v_- \in \Delta_+ + \Delta_-$ we obtain

$$\ell(g \cdot x)v = \ell_0(g) \cdot (1 + e_- \cdot x)(v_+ + v_-) = \ell_0(g)v_+ + \ell_0(g)v_- - \sqrt{2}\ell_0(g) \cdot x \cdot e_{n-1} \cdot v_+.$$

Comparing the components we see that $\ell(g \cdot x)v = v$ is equivalent to

$$\ell_0(g)v_+ = v_+ \quad \text{and} \quad \ell_0(g)v_- - \sqrt{2}\ell_0(g) \cdot x \cdot e_{n-1} \cdot v_+ = v_-.$$

Since the second condition holds for all $g \in G$, setting $g = 1$ we obtain $x \cdot e_{n-1} \cdot v_+ = 0$ for all $x \in \mathbb{R}^{n-2}$, which implies $v_+ = 0$ and $\ell_0(g)v_- = v_-$. Hence $\ell(H)v = v$ is equivalent to $v = v_-$ and $\ell_0(G)v_- = v_-$. □

These preparations allow us to prove the following result about the holonomy groups of Lorentzian spin manifolds with parallel spinor fields.

Theorem 16 *Let (M, g) be an oriented and time-oriented, indecomposable Lorentzian manifold of dimension n. Then (M, g) admits a spin structure with a non-zero parallel spinor field if and only if*

$$\mathrm{Hol}(M, g) \simeq G \ltimes \mathbb{R}^{n-2},$$

where $G \subset \mathrm{SO}(n-2)$ is a subgroup such that there is a non-zero fixed spinor in Δ_{n-2} under G. Moreover, the number N of linearly independent parallel spinor fields coincides with the dimension of the space of fixed spinors of Δ_{n-2} under G.

If $N > 0$, the connected component G^0 of G is a product of groups of the form $\{1\}$, $\mathrm{SU}(m)$, $\mathrm{Sp}(k)$, G_2 *or* $\mathrm{Spin}(7)$ *(with their standard representations). If G^0 acts irreducible on $\mathbb{R}^{n-2}$, then G is one of the groups in the table of Lemma 2.*

Proof Since (M, g) is indecomposable, by Corollary 2 each non-vanishing parallel spinor field has to be lightlike. Hence the full holonomy group is contained in the stabilizer of a lightlike vector, i.e. $\mathrm{Hol}(M, g) \subset \mathrm{SO}(n-2) \ltimes \mathbb{R}^{n-2}$. In particular, the holonomy algebra is of type 2 or 4 in Theorem 13. But, in case of type 4, the orthogonal part $\mathfrak{g}$ of $\mathfrak{hol}(M, g)$ is a holonomy algebra of a Riemannian manifold with a non-trivial center which excludes the existence of a fixed spinor in Δ_{n-2} under $\mathfrak{g}$ (cf. [60, 61]). Hence, the holonomy algebra $\mathfrak{hol}(M, g)$ is of type 2. In [15, Proposition 1] the possible structure of the full holonomy group for each of the holonomy algebra types is described. In particular, if the holonomy algebra $\mathfrak{hol}(M, g)$ is of type 2 then the full holonomy group $\mathrm{Hol}(M, g)$ has the form $\hat{G} \ltimes \mathbb{R}^{n-2}$, where $\hat{G} \subset \mathbb{R}^* \times \mathrm{O}(n-2)$ with connected component equal to the connected component of $\mathrm{proj}_{\mathrm{O}(n-2)}(Hol(M, g)) \subset \mathrm{O}(n-2)$. Since, in addition, $\mathrm{proj}_{\mathbb{R}^*}(Hol(M, g)) = \{1\}$, it follows that $\mathrm{Hol}(M, g) = G \times \mathbb{R}^{n-2}$, where $G \subset \mathrm{SO}(n-2)$. Hence, we can apply Proposition 5 and Lemma 3 and obtain the first part of the Theorem. The second part follows from Lemma 2. □

It remains to discuss whether there exist Lorentzian manifolds with the holonomy groups described in Theorem 16, where we are also interested to obtain manifolds with special causality properties.

Let us first consider again the construction in Example 6 and choose the function f there as independent on v. As we know, the Lorentzian manifold $M := \mathbb{R}^2 \times N$ with the metric $g^{f,h} := 2dvdu + fdu^2 + h$ has full holonomy group $\mathrm{Hol}(N, h) \ltimes \mathbb{R}^{n-2}$. For Lorentzian manifolds of type $(M, g^{f,h})$ various causality properties are known (see for example [30] and [37]). Let us quote here the following two results.

1) If (N, h) is a complete Riemannian manifold and if the function f does not depend on u and is at most quadratic at spacial infinity, i.e., there exist $x_0 \in N$ and real constants $r, c > 0$ such that

$$f(x) \leq c \cdot d_N(x_0, x)^2 \quad \text{for all } x \in N \text{ with } d_N(x_0, x) \geq r,$$

then $(M, g^{f,h})$ is geodesically complete. Here d_N is the distance function of (N, h).

2) If (N, h) is a complete Riemannian manifold and if the function $-f$ is spacial subquadratic, i.e., there exist $x_0 \in N$ and continious functions $p, c_1, c_2 \in C(\mathbb{R}, [0, \infty))$ with $p(u) < 2$ such that

$$-f(u, x) \leq c_1(u)\, d_N(x_0, x)^{p(u)} + c_2(u) \quad \text{for all } (u, x) \in \mathbb{R} \times N,$$

then $(M, g^{f,h})$ is globally hyperbolic.

Of course, both conditions for f can be realised in addition to $\det\big(\mathrm{Hess}^N f(u_0, x_0)\big) \neq 0$. Hence, each of the groups in Theorem 16 can be realised as holonomy group of a geodesically complete as well as by a globally hyperbolic Lorentzian manifold, if the group G in this Theorem is the holonomy group of a *complete* Riemannian manifold.

Another construction of globally hyperbolic examples in the form of Theorem 2 with complete Cauchy surface can be found in [13] and [15]. We consider the following warped product (F, g_F) over an irreducible Riemannian spin manifold (N^{n-2}, h) with parallel spinor fields:

$$F := \mathbb{R} \times N, \qquad g_F := ds^2 + e^{-4s} h.$$

Let $C : TF \to TF$ be a Codazzi tensor on (F, g_F), i.e. a symmetric homomorphism field such that

$$d^{\nabla^F} C(X, Y) := (\nabla^F_X C)Y - (\nabla^F_Y C)X = 0 \quad \text{for all } X, Y \in \mathfrak{X}(F),$$

with only positive eigenvalues and let $a \in \mathbb{R}$ be a positive number. Then the Lorentzian manifold $(M, g^{h,C})$,

$$M := (-a, \infty) \times F, \quad g^{h,C} := -dt^2 + g_t := -dt^2 + (C + 2(t + a)\mathrm{Id}_{TF})^* g_F,$$

has full holonomy $Hol(N, h) \ltimes \mathbb{R}^{n-2}$, hence it admits parallel spinor fields. If, in addition, (N, h) is complete, then the Lorentzian manifold $(M, g^{h,C})$ is globally hyperbolic and the space-like slices $(\{t\} \times F, g_t)$ are complete Cauchy hypersurfaces. Codazzi tensors on warped products are described in [13].

These two constructions reduce the problem of finding, for each G in Theorem 16, a Lorentzian manifold with holonomy $G \ltimes \mathbb{R}^{n-2}$ to the Riemannian case. First, one has to ensure the existence of Riemannian manifolds with holonomy group G. Then, for geodesically complete or globally hyperbolic Lorentzian metrics, one needs complete Riemannian manifolds with holonomy group G. For connected holonomy groups $G = G^0$ there are deep existence results for complete and even compact Riemannian manifolds with special holonomy obtained by several authors (for an overview see [52]). Based on the examples with connected holonomy groups, Moroianu and Semmelmann in [67] constructed irreducible Riemannian manifolds with parallel spinor for each of the non-connected groups G in the table of Lemma 2. For $\mathrm{SU}(m) \rtimes \mathbb{Z}_2$ they construct a compact manifold, and for the remaining groups the metrics are obtained by removing points from compact spaces or by cone constructions, thus these metrics are not complete. This yields the following conclusion.

Corollary 3 *For every group G in Theorem 16, which is connected or for which G^0 acts irreducible on $\mathbb{R}^{n-2}$, there exist a Lorentzian spin manifold with holonomy $G \ltimes \mathbb{R}^{n-2}$ and a non-zero parallel spinor field. Moreover, for the connected groups G and for $\mathrm{SU}(m) \rtimes \mathbb{Z}_2$, there exist geodesically complete as well as globally hyperbolic Lorentzian manifolds with complete spacelike Cauchy hypersurfaces and holonomy $G \ltimes \mathbb{R}^{n-2}$.*

To our knowledge, it remains an *open question* whether the groups $\mathrm{Sp}(m) \times \mathbb{Z}_d$, and $\mathrm{Sp}(m) \cdot \Gamma$ in the table of Lemma 2 can be realised as holonomy groups of *complete* Riemannian manifolds.

5 Constraint Equations for Special Lorentzian Holonomy

In the remaining half of this article, we will describe how one can construct (globally hyperbolic) Lorentzian manifolds with special holonomy by solving a Cauchy problem with initial conditions along a space-like hypersurface (see Sect. 7). As a starting point, in this section we will derive the constraint conditions special holonomy induces on spacelike hypersurfaces. We will mainly restrict to the case that the Lorentzian manifold admits a lightlike parallel vector field or a lightlike parallel spinor field.

5.1 Constraint Equations for Recurrent and Parallel Vector Fields

Let (M, g) be a time-oriented Lorentzian manifold and $i : \Sigma \hookrightarrow M$ a spacelike hypersurface with the induced Riemannian metric $h := i^*g$. For any $x \in \Sigma$

$$T_xM = T_x\Sigma \oplus \mathbb{R}T(x),$$

where $T : \Sigma \to TM|_\Sigma$ is the future-directed unit normal vector field on Σ. Then for all vector fields X, Y on Σ, we can decompose the covariant derivative $\nabla^g_X Y$ in a tangent and a normal part and obtain

$$\nabla^g_x Y \;=\; \mathrm{proj}_{T\Sigma}\nabla^g_X Y - g(\nabla^g_X Y, T)\,T \;= \nabla^h_X Y - \mathrm{II}(X, Y) \cdot T, \tag{22}$$

where II is the second fundamental form of the hypersurface $\Sigma \subset M$. If we denote by $W : T\Sigma \to T\Sigma$ the Weingarten operator of $\Sigma \subset M$,

$$W(X) := -\nabla^g_X T,$$

then the second fundamental form is given as

$$\mathrm{II}(X, Y) = g(W(X), Y) = h(W(X), Y).$$

Now we derive the constraint equations which a recurrent and lightlike vector field on a Lorentzian manifold imposes on a spacelike hypersurface.

Proposition 6 *Let (M, g) be a time-oriented Lorentzian manifold with a recurrent, lightlike, future-directed vector field V and let $\Sigma \subset M$ be a spacelike hypersurface with the induced Riemannian metric h and the future-directed unit normal vector field T. Then the function u and the vector field U on Σ, defined by*

$$u := -g(T, V|_\Sigma), \qquad U := uT - V|_\Sigma = -\mathrm{pr}_{T\Sigma}(V|_\Sigma),$$

satisfy

$$\nabla^h U = -u\, W + (\omega|_{T\Sigma}) \otimes U, \tag{23}$$

$$u^2 = h(U, U), \tag{24}$$

where ω is the 1-form on M defined by $\nabla^g V = \omega \otimes V$. In particular, $dU^\flat = \omega \wedge U^\flat$ and hence the distribution $U^\perp$ on Σ is integrable.

In particular, if V is parallel, U satisfies the constraints (24) and

$$\nabla^h U = -u\, W, \tag{25}$$

and $U^\flat$ is a nowhere vanishing closed 1-form on Σ.

Proof By definition we have

$$h(U, U) = g(uT - V|_\Sigma, uT - V|_\Sigma) = -2ug(V|_\Sigma, T) + u^2 g(T, T) = u^2.$$

Moreover, since V is recurrent, for all vector fields X on Σ holds

$$\begin{aligned} \omega(X)V = \nabla^g_X V &= \nabla^g_X(uT - U) \\ &= X(u)T + u\nabla^g_X T - \nabla^g_X U \\ &= X(u)T - uW(X) - \nabla^h_X U + h(W(U), X)T, \end{aligned}$$

which is equivalent to

$$\begin{aligned} X(u) &= -h(W(U), X) + u\omega(X) \qquad \text{and} \\ \nabla^h_X U &= -uW(X) + \omega(X)U, \end{aligned}$$

where the first condition follows from the second by differentiating the equation $u^2 = h(U, U)$. Then, since the Weingarten operator is symmetric, we obtain

$$dU^b(X, Y) = \tfrac{1}{2}\left(\omega(X)h(U, Y) - \omega(X)h(U, Y)\right).$$

This implies that the distribution $U^\perp$ is integrable. Moreover, the result for V parallel follows from setting $\omega = 0$. □

We will now describe the local geometry of Riemannian manifold (Σ, h) satisfying the constraint equations. We will restrict to the case where V is parallel. Here admitting a closed non-vanishing 1-form has strong implication for the local form of h as well as on the global geometry of Σ. For a proof see [62].

Theorem 17 *Any Riemannian manifold (Σ, h) that solves the constraints (24) and (25) with nowhere vanishing vector field U is locally isometric to*

$$\left(I \times N, h = u^{-2}ds^2 + g_s\right), \tag{26}$$

where $I \subset \mathbb{R}$ is an interval, g_s is a family of Riemannian metrics on some manifold N parametrised by $s \in I$, and $u^2 = h(U, U)$. Under this isometry, we have $U_s = u_s^2\partial_s$ and in the decomposition $T\Sigma = \mathbb{R}\partial_s \oplus TN$ the endomorphism W is given by

$$W = -\frac{1}{u}h(\nabla U, \cdot) = \begin{pmatrix} \partial_s(\frac{1}{u}) & grad^{g_s}(\frac{1}{u(s,.)}) \\ d(\frac{1}{u(s,.)}) & -\frac{u}{2}\mathcal{L}_{\partial_s} g_s \end{pmatrix}, \tag{27}$$

where $\mathcal{L}_{\partial_s}$ denotes the Lie derivative in s-direction, grad^{g_s} *the gradient with respect to g_s, and d the differential in N-directions.*

Moreover, if the vector field $\frac{1}{u^2}U$ is complete, then the universal cover of Σ is globally isometric to a manifold of the form (26) with $I = \mathbb{R}$ and N simply connected.

Conversely, given (Σ, h) as in (26) with $I = \mathbb{R}$ or $I = S^1$ a circle, the vector field $U = u^2\partial_s$ solves (25) for W as in (27). If in addition N is compact and u bounded, then (Σ, h) is complete.

The *proof* in [62] proceeds as follows:

Since $U^\flat$ is closed, locally, U is the gradient of some function z and the leaves of $U^\perp$ are given by its level sets $N_c = \{z = c\}$. For the vector field $Z = \frac{1}{h(U,U)}U$ we have

$$\mathcal{L}_Z U^\flat = dU^\flat(Z, .) = 0,$$

and hence, the flow ϕ of Z sends level sets of z to level sets, i.e., $\phi_s(p) \in N_{z(p)+s}$. Then the map

$$\Psi : I \times N_0 \ni (s, p) \mapsto \phi_s(p),$$

is a local diffeomorphism with $d\Psi(\partial_s) = Z$ and which, since its differential preserves $U^\perp$, pulls back the metric to

$$\Psi^* h = u^{-2} ds^2 + g_s.$$

The statement about the universal covering space is based on the following fact, for a proof see for example [63, Proposition 8].

Lemma 4 *Let Σ be a smooth manifold with a closed 1-form η and a complete vector field Z such that $\eta(Z) = 1$. Then all leaves N of $\ker(\eta)$ are diffeomorphic to each other and the universal cover $\widetilde{\Sigma}$ of Σ is diffeomorphic to $\mathbb{R} \times \widetilde{N}$, where $\widetilde{N}$ is the universal cover of N.*

The above Theorem 17 also give us a way to construct *complete* Riemannian manifolds satisfying the constraint conditions.

5.2 Constraint Equations for Parallel Spinor Fields

Next, let us suppose in addition, that (M, g) is a Lorentzian spin manifold of dimension $(n + 1)$, with spin structure $(\widetilde{\mathscr{F}}(M, g), \Lambda)$. Then each spacelike hypersurface Σ is oriented by a timelike future directed unit normal T and $(\widetilde{\mathscr{F}}(M, g), \Lambda)$ induces a spin structure on (Σ, h) as follows. The embedding of the frame-bundle

$$\begin{aligned} \mathscr{F}(\Sigma, h) &\longrightarrow \mathscr{F}(M, g)|_\Sigma \\ (s_1, \ldots, s_n) &\longmapsto (T, s_1, \ldots, s_n) \end{aligned}$$

is a reduction of $\mathscr{F}(M, g)|_\Sigma$ to the subgroup $\mathrm{SO}(n) \subset \mathrm{SO}^0(1, n)$. Hence $\widetilde{\mathscr{F}}(\Sigma, h) := \Lambda^{-1}(\mathscr{F}(\Sigma, h)) \subset \widetilde{\mathscr{F}}(M, g)|_\Sigma$ is a principal fibre bundle with structure group $\mathrm{Spin}(n)$ and gives together with the map $\Lambda_\Sigma := \Lambda|_{\widetilde{\mathscr{F}}(\Sigma,h)}$ a spin structure of (Σ, h). If we restrict the spin representation $\Delta_{1,n}$ to the subgroup $\mathrm{Spin}(n) \subset \mathrm{Spin}^0(1, n)$, we obtain the following isomorphisms of $\mathrm{Spin}(n)$-representations:

$$\Delta_n \simeq \begin{cases} \Delta_{1,n}|_{\mathrm{Spin}(n)}, & \text{if } n \text{ is even}, \\ \Delta_{1,n}^+|_{\mathrm{Spin}(n)}, & \text{if } n \text{ is odd}. \end{cases}$$

Hence, the spinor bundle S^Σ of (Σ, h) and the spinor bundle S of (M, g) are related by

$$\begin{aligned} S^\Sigma = \widetilde{\mathscr{F}}(\Sigma, h) \times_{\mathrm{Spin}(n)} \Delta_n &= \begin{cases} \widetilde{\mathscr{F}}(M, g)|_\Sigma \times_{\mathrm{Spin}^0(1,n)} \Delta_{1,n}, & \text{if } n \text{ is even}, \\ \widetilde{\mathscr{F}}(M, g)|_\Sigma \times_{\mathrm{Spin}^0(1,n)} \Delta_{1,n}^+, & \text{if } n \text{ is odd}. \end{cases} \\ &= \begin{cases} S|_\Sigma & \text{if } n \text{ is even} \\ S^+|_\Sigma & \text{if } n \text{ is odd}. \end{cases} \end{aligned} \tag{28}$$

Let $\phi \in \Gamma(S^{(+)})$ be a spinor field on (M, g) and $\varphi := \phi|_\Sigma \in \Gamma(S^\Sigma)$ the corresponding spinor field on (Σ, h) given by the identification (28). Then the Clifford multiplication in the two spinor bundles is related by

$$X \cdot \varphi = i\, T \cdot X \cdot \phi|_\Sigma \qquad \text{for all } X \in \mathfrak{X}(\Sigma) \tag{29}$$

and the positive definite scalar products on both bundles satisfy

$$\langle \varphi_1, \varphi_2 \rangle_{S^\Sigma} = (\phi_1|_\Sigma, \phi_2|_\Sigma)_T. \tag{30}$$

The next proposition describes the constraint conditions which a lightlike parallel spinor field on a Lorentzian spin manifold imposes on a spacelike hypersurface.

Proposition 7 *Let (M, g) be a Lorentzian spin manifold, $\Sigma \subset M$ a spacelike hypersurface with a future directed timelike unit normal field T and $\phi \in \Gamma(S^{(+)})$ a lightlike parallel spinor field on (M, g). Then the induced spinor field $\varphi := \phi|_\Sigma$ on (Σ, h) satisfies*

$$\nabla^{S^\Sigma}_X \varphi = \frac{i}{2} W(X) \cdot \varphi \qquad \textit{for all } X \in \mathfrak{X}(\Sigma), \tag{31}$$

$$U_\varphi \cdot \varphi = i\, u_\varphi \varphi, \tag{32}$$

where W denotes the Weingarten operator of the hypersurface $\Sigma \subset M$, U_φ is the Dirac current of φ and u_φ is the positive function given by

$$u_\varphi = \|U_\varphi\|_h = \|\varphi\|^2.$$

Moreover, $U_\varphi = \mathrm{proj}_{T\Sigma}(-V_\phi|_\Sigma)$.

Proof Let $(s_1, \ldots, s_n)$ be a local orthonormal frame on Σ and $X \in \mathfrak{X}(\Sigma)$. From (5) we obtain for the spinor derivative of ϕ:

$$\begin{aligned}\nabla^S_X \phi &= X(\phi) - \frac{1}{2}\sum_{j=1}^n g(\nabla^g_X T, s_j) T \cdot s_j \cdot \phi + \frac{1}{2}\sum_{k<j} g(\nabla^g_X s_k, s_j) s_k \cdot s_j \cdot \phi \\ &= X(\phi) + \frac{i}{2}\sum_{j=1}^n g(\nabla^g_X T, s_j)(iT \cdot s_j) \cdot \phi + \frac{1}{2}\sum_{k<j} g(\nabla^g_X s_k, s_j)(iT \cdot s_k) \cdot (iT \cdot s_j) \cdot \phi.\end{aligned}$$

The identification of the spinor bundles and formula (29) for the Clifford multiplication imply

$$\begin{aligned}\nabla^S_X \phi|_\Sigma &= X(\varphi) - \frac{i}{2}\sum_{j=1}^n h(W(X), s_j) s_j \cdot \varphi + \frac{1}{2}\sum_{k<j} h(\nabla^h_x s_k, s_j) s_k \cdot s_j \cdot \varphi \\ &= \nabla^{S^\Sigma}_X \varphi - \frac{i}{2} W(X) \cdot \varphi.\end{aligned}$$

This shows (31). To derive the algebraic condition (32), we decompose the Dirac current V_φ of ϕ on Σ into $V_\phi|_\Sigma = uT - U$, where $U = \mathrm{proj}_{T\Sigma}(-V_\phi|_\Sigma)$, and use the condition $V_\phi \cdot \phi = 0$ (cf. Proposition 3). Multiplying $0 = (uT - U)\cdot\phi)|_\Sigma$ with T yields

$$0 = (uT \cdot T \cdot \phi)|_\Sigma + i(iT \cdot U)\cdot\phi|_\Sigma \;=\; u\varphi + iU\cdot\varphi.$$

It remains to prove that U is the Dirac current U_φ of φ and $u = \|\varphi\|^2$. By definition,

$$h(U_\varphi, X) = -i\langle X\cdot\varphi, \varphi\rangle_{S^\Sigma}.$$

On the other hand, for all $X \in \mathfrak{X}(\Sigma)$,

$$\begin{aligned} h(U,X) \;&=\; -g(V_\phi|_\Sigma, X) \;=\; \langle X\cdot\phi, \phi\rangle|_\Sigma = (T\cdot X\cdot\phi, \phi)_T \\ &\overset{(29),(30)}{=} -i\langle X\cdot\varphi,\varphi\rangle_{S^\Sigma} \;=\; h(U_\varphi, X). \end{aligned}$$

Hence $U_\varphi = U = \mathrm{proj}(-V_\phi|_\Sigma)$. Moreover, by Proposition 6,

$$h(U_\varphi, U_\varphi) = u^2 = -i\langle U_\varphi\cdot\varphi,\varphi\rangle_{S^\Sigma} = -i^2u\langle\varphi,\varphi\rangle_{S^\Sigma} = u\|\varphi\|^2,$$

which finishes the proof. □

Definition 9 Let (Σ, h) be a Riemannian spin manifold and $W : T\Sigma \to T\Sigma$ a symmetric endomorphism field. A spinor field $\varphi \in \Gamma(S^\Sigma)$ is called *imaginary W-Killing spinor* or *imaginary generalised Killing spinor*, if it satisfies

$$\nabla_X^{S^\Sigma}\varphi = \frac{i}{2}W(X)\cdot\varphi \qquad \text{for all } X \in \mathfrak{X}(\Sigma).$$

This definition relates to the constraint conditions for a parallel lightlike vector field as follows.

Lemma 5 *Let (Σ, h) be a Riemannian manifold with an imaginary W-Killing spinor φ. Then the Dirac current U_φ satisfies the following conditions*

1. $\nabla^h U_\varphi = -\|\varphi\|^2 \cdot W$.
2. *The function $q_\varphi := \|\varphi\|^4 - h(U_\varphi, U_\varphi)$ is non-negative and constant. Moreover*

$$q_\varphi = \|\varphi\|^2 \cdot \mathrm{dist}(i\varphi, E_\varphi)^2,$$

where $\mathrm{dist}(i\varphi, E_\varphi)$ is the pointwise distance of the spinor $i\varphi$ to the subspace $E_\varphi := \{X\cdot\varphi \mid X \in TM\} \subset S_\Sigma$ with respect to the real scalar product $Re\langle\cdot,\cdot\rangle_{S_\Sigma}$.

If $q_\varphi = 0$, then $u_\varphi := \|\varphi\|^2 = \|U_\varphi\|_h$ and $U_\varphi\cdot\varphi = iu_\varphi\varphi$. In particular, in this case U_φ satisfies the the constraint conditions (24) and (25) with respect to W.

Proof Differentiating $h(U_\varphi, X) = i\langle\varphi, X\cdot\varphi\rangle$ yields

$$\begin{aligned} h(\nabla^h_Y U_\varphi, X) &= -\tfrac{1}{2}\langle W(Y)\cdot\varphi, X\cdot\varphi\rangle + \tfrac{1}{2}\langle\varphi, X\cdot W(Y)\cdot\varphi\rangle \\ &= \tfrac{1}{2}\langle\varphi, (W(Y)\cdot X + X\cdot W(Y))\cdot\varphi\rangle \\ &= -\|\varphi\|^2\cdot h(W(Y), X). \end{aligned}$$

This shows the first equation. Furthermore,

$$\begin{aligned} Y(\|\varphi\|^4) &= 2\|\varphi\|^2\cdot(\langle\nabla_Y\varphi,\varphi\rangle + \langle\varphi,\nabla_Y\varphi\rangle) \\ &= \|\varphi\|^2\cdot(i\langle W(Y)\cdot\varphi,\varphi\rangle - i\langle\varphi, W(Y)\cdot\varphi\rangle) \\ &= -2\|\varphi\|^2\cdot h(U_\varphi, W(Y)). \end{aligned}$$

Hence,

$$Y(q_\varphi) = -2\|\varphi\|^2\cdot h(U_\varphi, W(Y)) + 2\|\varphi\|^2\cdot h(U_\varphi, W(Y)) = 0,$$

i.e., the function q_φ is constant. In order to show, that q_φ is non-negative, we calculate for the distance $\mathrm{dist}(i\varphi, E_\varphi)$:

$$\begin{aligned} \mathrm{dist}(i\varphi, E_\varphi)^2 &= \|i\varphi\|^2 - \|\mathrm{proj}_{E_\varphi}(i\varphi)\|^2 = \|\varphi\|^2 - \|\varphi\|^{-4}\cdot\|\sum_j\langle i\varphi, s_j\cdot\varphi\rangle\, s_j\cdot\varphi\|^2 \\ &= \|\varphi\|^2 - \|\varphi\|^{-2}\cdot h(U_\varphi, U_\varphi) \;=\; q_\varphi\cdot\|\varphi\|^{-2}. \end{aligned}$$

In the case of $q_\varphi = 0$ there exists a vector field ξ on Σ such that $\xi\cdot\varphi = i\varphi$. Using the definition of U_φ shows that $\xi = \|\varphi\|^{-2}\cdot U_\varphi$ and therefore, $U_\varphi\cdot\varphi = iu_\varphi\varphi$. □

Definition 10 We say that an imaginary W-Killing spinor is of type I if $q_\varphi = 0$, and of type II if $q_\varphi > 0$.

Remark 3 The restriction of a parallel spinor field ϕ of a Lorentzian manifold to a spacelike hypersurface is always an imaginary W-Killing spinor. By Proposition 7 $\varphi := \phi|_\Sigma$ is of type I, if V_ϕ is lightlike. The same arguments as in proof of this proposition shows that φ is of type II, if V_ϕ it timelike.

Remark 4 (Killing Spinors) As the name suggests, imaginary generalised Killing spinors are a generalisation of *imaginary Killing spinors* for which $W = 2\mu\mathrm{Id}$ for $\mu\in\mathbb{R}\setminus\{0\}$. These are, in turn, a special case of Riemannian manifolds (Σ, h) with Killing spinors, i.e., with $\nabla^h\varphi = \lambda X\cdot\varphi$ with $\lambda\in\mathbb{C}\setminus\{0\}$. The existence of a Killing spinor on (Σ, h) of dimension n implies that h is an Einstein metric with scalar curvature $4n(n-1)\lambda^2$, which in particular implies that λ is either real or imaginary. By Myer's Theorem, it implies that complete Riemannian manifolds with real Killing spinor ($\lambda\in\mathbb{R}\setminus\{0\}$) are compact. On the other hand if (Σ, h) admits an imaginary Killing spinor ($\lambda = i\mu\in i\mathbb{R}\setminus\{0\}$), Σ cannot be compact, as otherwise the inequality

$$0 \leq \int_{\Sigma} \langle D\varphi, D\varphi \rangle d\Sigma = \int_{\Sigma} \langle D^2\varphi, \varphi \rangle d\Sigma = -n\mu^2 \int_{\Sigma} \langle \varphi, \varphi \rangle d\Sigma < 0,$$

would hold, in which D is the Dirac operator. In contrast, Riemannian manifolds with imaginary *generalised* Killing spinors can be compact, see [16, Section 6] for a 2-dimensional example.

Moreover, for a Riemannian manifold with a Killing spinor, the (Riemannian or Lorentzian) cone

$$M = \mathbb{R}_{>0} \times \Sigma, \quad g = 2\lambda^2 dr^2 + r^2 h,$$

admits a parallel spinor.[5] This, together with Berger's classification, Wang's result ([75], as described in our Proposition 5) and a result by Gallot that a cone over a complete Riemannian manifold is either flat or irreducible [47], was used by C. Bär in [4] when classifying complete Riemannian manifolds with real Killing spinors. Generalised real Killing spinors have been considered in [3].

The first author obtained the following classification of complete Riemannian manifolds with imaginary Killing spinor φ in [9]: if φ is of type II, then (Σ, h) is isometric to the hyperbolic space with sectional curvature $-4\mu^2$, and if φ is of type I, then (Σ, h) is isometric to a warped product of the form

$$(\mathbb{R} \times N,\ ds^2 + \mathrm{e}^{-4\mu s} g),$$

where (N, g) is a complete Riemannian spin manifold with parallel spinor field. This result can also be deduced using a generalisation of Gallot's result in [2].

Since for a Riemannian manifold (Σ, h) with imaginary generalised W-Killing spinor φ of type I the vector field U_φ satisfies the constraint conditions (25), Theorem 17 applies to (Σ, h) and consequently leads to Theorem 28 in Sect. 8.2, which can be seen as a (local) analogue of the classification for imaginary Killing spinors in [9].

For the moment, let us present classification results for Riemannian manifolds with imaginary generalised W-Killing spinors for two special cases of W.

Example 8 $W := f \cdot \mathrm{Id}_{T\Sigma}$ with $f \in C^\infty(\Sigma, \mathbb{R}^+)$.

Then (Σ, h) is a complete Riemannian spin manifold with imaginary W-Killing spinor of type I if and only if (Σ, h) is isometric to a warped product

$$(\mathbb{R} \times F\,,\ ds^2 + e^{-4\int_0^s f(\tau)\, d\tau} g_F),$$

[5] This is in fact true in any signature.

where (F, g_F) is a complete Riemannian spin manifold with a non-vanishing parallel spinor field (for a proof see [14, chap.7.2] or [69]).

In this example, F is a level set of the function $u_\varphi = \|\varphi\|^2$ and the splitting diffeomorphism is given by the flow of the vector field $u_\varphi^{-1} U_\varphi$. In this spitting, the function f only depends on the s-coordinate.

Example 9 Let W be a Codazzi tensor on (Σ, h), i.e.

$$d^{\nabla^h} W(X, Y) := (\nabla^h_X W)(Y) - (\nabla^h_Y W)(X) = 0 \qquad \text{for all } X, Y \in \mathfrak{X}(\Sigma),$$

and suppose that W is invertible and all of its eigenvalues λ are positive and uniformally bounded by positive constants, $0 < c_1 \leq \lambda \leq c_2$.

In this case, (Σ, h) is a complete Riemannian spin manifold with imaginary W-Killing spinor of type I if and only if (Σ, h) is isometric to

$$(\mathbb{R} \times F\,,\ (W^{-1})^*(ds^2 + e^{-4s} g_F)\,),$$

where (F, g_F) is a complete Riemannian spin manifold with a non-vanishing parallel spinor field and W^{-1} is a Codazzi tensor for the warped product $(\mathbb{R} \times F, ds^2 + e^{-4s} g_F)$. In [13] Codazzi tensors on such warped products are described.

6 The Cauchy Problem for the Vacuum Einstein Equations

The results in the previous section suggest the question whether the constraint conditions for a Riemannian manifold (Σ, h) are not only necessary but also sufficient for (Σ, h) being a Cauchy hypersurface in a Lorentzian manifolds with parallel lightlike vector field (or spinor field). In Sect. 7 we will see that this is indeed the case by showing how a globally hyperbolic Lorentzian manifold with special holonomy can be constructed from a given Riemannian manifold (Σ, h) satisfying the constraint equations in Sect. 5. We will achieve this by studying certain Cauchy problems for PDEs that are induced by the existence of parallel lightlike vector fields or spinor fields. Since our approach for establishing the existence of a solution is in parts analogous to the approach for the Einstein equations, in this section we will review this approach in detail. We will however focus on local aspects of the proofs, i.e., on constructing a neighbourhood of a given Riemannian manifold, and not on much more difficult aspects such as maximality or stability of solutions.

6.1 The Constraint Conditions for the Vacuum Einstein Equations

The vacuum Einstein equations for a Lorentzian manifold (M, g) are equivalent to the condition that g is Ricci-flat,

$$Ric^g = 0. \tag{33}$$

In this equation the Lorentzian metric g is considered to be the unknown and the equation for g is usually treated as a Cauchy problem: given a Riemannian manifold (Σ, h), is there a Lorentzian manifold (M, g) satisfying equation (33) such that (Σ, h) is a spacelike hypersurface in (M, g) and $g|_\Sigma = h$? Of course, for this to be possible, (Σ, h) has to satisfy certain *constraint equations*, which we will now derive.

Let (M, g) be a time oriented Lorentzian manifold and $(\Sigma, h = g|_\Sigma)$ be a spacelike hypersurface with Weingarten tensor W and second fundamental form II. Then the curvature tensors of g and h satisfy the *Gauss equation*,

$$R^g(X, Y, Z, V) = R^h(X, Y, Z, V) + \mathrm{II}(Y, Z)\mathrm{II}(X, V) - \mathrm{II}(Y, V)\mathrm{II}(X, Z), \tag{34}$$

and the Codazzi equation,

$$R^g(X, Y, Z, T)|_{T\Sigma} = (\nabla^h_X \mathrm{II})(Y, Z) - (\nabla^h_Y \mathrm{II})(X, Z), \tag{35}$$

where X, Y, Z, V vectors tangent to Σ and T is the positive timelike unit normal along Σ. They follow from the formula (22) and imply for the Ricci tensor of g that

$$Ric^g(T, X) = \mathrm{tr}(\nabla^h_X W) - \mathrm{div}^h(W)(X) = d\mathrm{tr}(W)(X) - \mathrm{div}^h(\mathrm{II})(X), \tag{36}$$

and

$$Ric^g(X, Y) = -R^g(T, X, Y, T) + Ric^h(X, Y) + \mathrm{tr}(W)\mathrm{II}(X, Y) - \mathrm{II}(W(X), Y), \tag{37}$$

again with $X, Y \in T\Sigma$. In addition we have that

$$Ric^g(T, T) = \sum_{i=1}^{n} R^g(T, E_i, E_i, T),$$

where $T, E_1, \dots E_n$ is an orthonormal basis for g. This implies for the scalar curvature that

$$scal^g = -2Ric^g(T, T) + scal^h + \mathrm{tr}(W)^2 - \mathrm{tr}(W^2). \tag{38}$$

Now one tries to express $Ric^g = 0$ along Σ in terms of geometric quantities in (Σ, h). The resulting equations, that are implied by the above formulae, are called the *constraint equations for the vacuum Einstein equations* (see for example [7]):

Lemma 6 (Constraint Equations for the Vacuum Einstein Equations) *Let (M, g) be a time oriented Lorentzian manifold with $Ric^g = 0$ and $(\Sigma, h = g|_\Sigma)$ be a spacelike hypersurface with unit normal T and Weingarten tensor W. Then $Ric^g(T, .)|_{T\Sigma} = 0$ is equivalent to*

$$d\mathrm{tr}(W) - \mathrm{div}^h(W) = 0. \tag{39}$$

and $Ric^g(T, T)|_\Sigma = 0$ is equivalent to

$$scal^g|_\Sigma = scal^h + \mathrm{tr}(W)^2 - \mathrm{tr}(W^2).$$

In particular, if $Ric^g = 0$ along Σ, then the constraint equations (39) and

$$scal^h + \mathrm{tr}(W)^2 - \mathrm{tr}(W^2) = 0, \tag{40}$$

are satisfied.

6.2 Results from PDE Theory

We will now state the results from the theory of PDEs that are needed in order to show the existence of solutions for the considered Cauchy problems. We will formulate them for functions on $\mathbb{R}^{n+1}$ with values in some $\mathbb{R}^N$. We fix coordinates $(x^0, \ldots, x^n)$ on $\mathbb{R}^{n+1}$ and use the index convention that Greek indices $\mu, \nu, \ldots$ run from 0 to n and Latin indices $i, j, \ldots$ only from 1 to n. The initial data will be given on a hypersurface Σ which we assume to be an open set in the hyperplane $\{x^0 = 0\}$ containing the origin.

The first result is the *Cauchy-Kowalevski Theorem* and we will formulate it for PDEs of second order. It gives an existence and uniqueness statement for a Cauchy problem of the following form: find an $\mathbb{R}^N$-valued function w, defined on a neighbourhood in $\mathbb{R}^{n+1}$ of the origin such that

$$\partial_0^2 w = F(x^\mu, w, \partial_\mu w, \partial_0\partial_i w, \partial_i\partial_j w), \qquad w|_\Sigma = f, \quad \partial_0 w|_\Sigma = g, \tag{41}$$

where f and g are functions on Σ and F is an R^N-valued function on a vector space of the appropriate dimension.

Theorem 18 (Cauchy-Kowalevski) *If F, f and g are real analytic, then the Cauchy problem (41) has a unique analytic solution w defined in a neighbourhood of the origin in $\mathbb{R}^{n+1}$.*

For a proof see for example [38] or [74, Section 16].

Of course, real analyticity is a strong assumption, but since Lewy's example it is well known that it cannot be removed. Hence, we will give another result from the theory of *first order quasilinear symmetric hyperbolic systems* which applies in the smooth setting.

Definition 11 A first order, quasilinear PDE of the form (in which we use Einstein's summation convention),

$$A^{\nu}(x^{\mu}, w)\partial_{\nu} w = b(x^{\mu}, w), \tag{42}$$

for an $\mathbb{R}^N$-valued function w on $\mathbb{R}^{n+1}$, is called *symmetric hyperbolic*, if

- the A^{ν}'s and b are functions of the $n+1$ coordinates and the N unknowns,
- the A^{ν} have values in the *symmetric* $N \times N$-matrices,
- b has values in $\mathbb{R}^N$, and
- $A^0 \geq c\mathrm{Id}$ for a positive constant c.

For first order, quasilinear symmetric hyperbolic PDEs there is the following result.

Theorem 19 (Existence and Uniqueness for Symmetric Hyperbolic PDEs) *Consider a symmetric hyperbolic system (42) with A^{μ} and b smooth functions and $f \in C^{\infty}(\Sigma)$. Then there is a smooth solution $w : M \to \mathbb{R}^N$ to (42) on an open neighbourhood M of Σ in $\mathbb{R} \times \Sigma$ with $w|_{\Sigma} = f$. Such a solution is unique in the sense that if $w_i : M_i \to \mathbb{R}^N$, $i = 1, 2$, are two smooth solutions on open neighbourhoods M_i in $\mathbb{R} \times \Sigma$ with $w_1|_{\Sigma} = w_2|_{\Sigma}$, then $w_1 = w_2$ on $M_1 \cap M_2$.*

For more details and proofs we refer to [74, Sections 16.1 and 16.2] or [42] and references therein.

The bad news is that neither of the two results can be applied directly to the PDE that arises when (33) is written in *arbitrary* coordinates. Indeed, in coordinates $(x^0, \ldots, x^n)$ on a Lorentzian manifold (M, g) with $g = g_{\mu\nu} dx^{\mu} dx^{\nu}$, the Ricci tensor $Ric^g = Ric^g_{\mu\nu} dx^{\mu} dx^{\nu}$ is given by

$$Ric^g_{\mu\nu} = -\tfrac{1}{2} g^{\alpha\beta} \partial_{\alpha}\partial_{\beta} g_{\mu\nu} + \partial_{(\mu}\Gamma_{\nu)} + G_{\mu\nu}, \tag{43}$$

where $\Gamma_{\mu} = g^{\alpha\beta}\Gamma_{\mu\alpha\beta}$ and $\Gamma_{\mu\alpha\beta} = \frac{1}{2}\left(\partial_{\alpha} g_{\beta\mu} + \partial_{\beta} g_{\alpha\mu} - \partial_{\mu} g_{\alpha\beta}\right)$ denote the Christoffel symbols of g, the round brackets denote the symmetrisation of indices, and $G_{\mu\nu} = G_{\mu\nu}(x^{\alpha}, g_{\alpha\beta}, \partial_{\gamma} g_{\alpha\beta})$ are lower order terms depending on the coordinates x^{α}, the functions $g_{\alpha\beta}$ and its first derivatives $\partial_{\gamma} g_{\alpha\beta}$. The term that in general prevents us from applying the Cauchy-Kowalevski Theorem or (when rewriting this as a first order PDE) Theorem 19 is the term

$$\partial_{(\mu}\Gamma_{\nu)} = \tfrac{1}{2} g^{\alpha\beta}\left(\partial_{\mu}\partial_{\alpha} g_{\nu\beta} + \partial_{\nu}\partial_{\alpha} g_{\mu\beta} - \partial_{\mu}\partial_{\nu} g_{\alpha\beta}\right) + LOTs. \tag{44}$$

The fundamental reason that the above results cannot be applied directly to Eq. (33) is its *diffeomorphism invariance*. Indeed, let g be a Ricci-flat Lorentzian metric on $\mathbb{R}^{n+1}$ and ϕ a diffeomorphism of $\mathbb{R}^{n+1}$ that restricts to the identity in a small neighbourhood of $\Sigma = \{x^0 = 0\}$. Then the metric $\phi^* g$, which is the pull-back of g by ϕ, is isometric to g and thus also Ricci-flat. Hence, its metric coefficients $(\phi^* g)_{\mu\nu} = \partial_\mu \phi^\alpha \partial_\nu \phi^\beta g_{\alpha\beta}$ are also solutions to the PDE (33) but, because of the assumption we have $\phi|_\Sigma = \mathrm{Id}$ and $d\phi|_\Sigma = \mathrm{Id}$ and thus the *same initial condition* along Σ. This however contradicts the uniqueness statements in the above results. Hence, in order to apply the PDE theory to Eq. (33), one has to break this diffeomorphism invariance. We will show two ways how this can be done.

6.3 The Vacuum Einstein Equations as Evolution Equations

Let (Σ, h) be a Riemannian manifold satisfying the constraint conditions in Lemma 6. The aim is to construct a globally hyperbolic Lorentzian manifold (M, g) that is Ricci-flat and contains (Σ, h) as a Cauchy hypersurface. Motivated by the Splitting Theorem 2, we will make an anstaz that M is an open neighbourhood of $\{0\} \times \Sigma$ in $\mathbb{R} \times \Sigma$ and the metric g is of the form

$$g = -\lambda^2 dt^2 + h_t, \tag{45}$$

where h_t is a family of Riemannian metrics with $h_0 = h$ and λ a positive function on $\mathbb{R} \times \Sigma$. We are now going to derive the Ricci-flatness equation as an evolution equation for the family of Riemannian metrics h_t. In regards to the remarks in Sect. 6.2, in this setting the diffeomorphism invariance is broken by only allowing coordinate transformations of Σ, i.e., diffeomorphism that are independent of t.

In M we consider the hypersurfaces $\Sigma_t = \{t\} \times \Sigma$ equipped with the metrics $h_t = g|_{T\Sigma_t \times T\Sigma_t}$. Their timelike unit normal is given by $T = \lambda^{-1}\partial_t$, their Weingarten tensor by $W_t = -\nabla^g T|_{T\Sigma_t}$, and their second fundamental forms II_t by

$$\mathrm{II}_t(X, Y) = -\frac{1}{2\lambda}(\mathscr{L}_{\partial_t} g)(X, Y) \qquad \text{for all } X, Y \in T\Sigma_t, \tag{46}$$

where $\mathscr{L}_{\partial_t}$ denotes the Lie derivative with respect to ∂_t. In this setting we introduce some notation. If A is an endomorphism and B a symmetric bilinear form on M with $A(T) = T \lrcorner B = 0$, then it holds that

$$(\mathscr{L}_{\partial_t} A)(T) = T \lrcorner (\mathscr{L}_{\partial_t} B) = 0,$$

where $\mathscr{L}$ denotes the Lie derivative. Hence, we introduce the notation

$$A' := \mathscr{L}_{\partial_t} A, \qquad B' := \mathscr{L}_{\partial_t} B.$$

Furthermore, if A and B are metric duals of each other, $B = g(A.,.) = h_t(A.,.)$, we get

$$B'(X, Y) - g(A'X, Y) = -2\lambda \mathrm{II}_t(AX, Y),$$

which by Eq. (46) becomes the familiar Leibniz formula

$$B'(X,Y) = h_t(A'X,Y) + h_t'(AX,Y). \tag{47}$$

In case of the second fundamental form and the Weingarten operator, this implies

$$\mathrm{II}_t'(X,Y) = h_t(W_t'(X),Y) - 2\lambda\, h_t(W_t^2(X),Y).$$

In addition to the Gauß and Codazzi equations, in this situation we can also express the curvature term $R^g(T,X,Y,T)$ in terms of (derivatives of) II_t and get an equation that is sometimes called *Mainardi equation*,

$$R^g(X,T,T,Y) = \mathrm{II}_t(X,W_t(Y)) + \tfrac{1}{\lambda}\left(\mathrm{II}_t'(X,Y) + \mathrm{Hess}^{h_t}(\lambda)(X,Y)\right). \tag{48}$$

On the one hand, we can insert this into Eq. (37) and get

$$Ric^g|_{T\Sigma_t \times T\Sigma_t} = Ric^{h_t} + \mathrm{tr}(W_t)\,\mathrm{II}_t - 2\mathrm{II}_t(.,W_t.) - \tfrac{1}{\lambda}\left(\mathrm{II}_t' + \mathrm{Hess}^{h_t}(\lambda)\right), \tag{49}$$

where $\mathrm{Hess}^{h_t}(\lambda) = \nabla^{h_t} d\lambda$ denotes the Hessian of λ with respect to the metric h_t. On the other hand, we get the equation

$$Ric^g(T,T) = \mathrm{tr}(W_t^2) + \tfrac{1}{\lambda}\left(\mathrm{tr}^{h_t}(\mathrm{II}_t') + \Delta_{h_t}(\lambda)\right), \tag{50}$$

where $\Delta_{h_t}(\lambda) = \mathrm{tr}^{h_t}(\mathrm{Hess}^{h_t}(\lambda))$ denotes the Laplacian of λ with respect to h_t. Note the subtle notional difference here that the trace of the bilinear form is taken with respect to h_t while the trace for an endomorphism is independent of the metric. This yields the expression for the scalar curvature

$$scal^g = scal^{h_t} + (\mathrm{tr}(W_t))^2 - 3\mathrm{tr}(W_t^2) - \tfrac{2}{\lambda}\left(\mathrm{tr}^{h_t}(\mathrm{II}_t') + \Delta_{h_t}(\lambda)\right). \tag{51}$$

Then we observe:

Lemma 7 *The Lorentzian metric g in (45) is Ricci flat if and only if the equation*

$$\mathrm{II}_t' = \lambda\left(Ric^{h_t} + \mathrm{tr}(W_t)\,\mathrm{II}_t - 2\mathrm{II}_t(.,W_t.)\right) - \mathrm{Hess}^{h_t}(\lambda), \tag{52}$$

as well as the constraint equations (39) and (40) are satisfied for each Riemannian metric h_t.

Proof The ‘only-if’ direction of this statement clearly follows from Lemma 6 and formula (49).

On the other hand, by formula (49), Eq. (52) implies that $Ric^g|_{T\Sigma_t\times T\Sigma_t} = 0$. By formula (36), the constraint equation (39) implies that $Ric^g(T,X) = 0$ for all $X \in T\Sigma_t$ and it remains to show that $Ric(T,T) = 0$. For this, we take the trace of

Eq. (52) and the constraint (40) to get

$$0 = \mathrm{tr}^{h_t}\mathrm{II}'_t + \Delta_{h_t}(\lambda) - \lambda\left(scal^{h_t} + (\mathrm{tr}(W_t))^2 - 2\mathrm{tr}(W_t^2)\right)$$
$$= \mathrm{tr}^{h_t}\mathrm{II}'_t + \Delta_{h_t}(\lambda) + \lambda\mathrm{tr}(W_t^2).$$

By (50) this implies that $Ric^g(T,T) = 0$ and hence $Ric^g = 0$. □

The key observation now is that Eq. (52) is a second order *evolution equation* for the family h_t of Riemannian metrics that it is of Cauchy-Kowalevski form as in (41). This becomes evident when spelling out the left-hand-side of (52) as

$$\mathrm{II}'_t = \tfrac{1}{2\lambda^2}\left(\lambda' h'_t - \lambda h''_t\right),$$

and noting that the right-hand-side of (52) contains no double derivatives of h_t in t-direction. Then the Cauchy-Kowalevski Theorem applies:

Theorem 20 *Let (Σ, h) be a real analytic Riemannian manifold satisfying the constraint equations (40) and (39) with a real analytic symmetric endomorphism W and corresponding symmetric bilinear form* II, *and let λ be a real analytic function on $\mathbb{R}\times\Sigma$. Then there is a unique real analytic Ricci-flat Lorentzian metric g of the form (45) with $h_0 = h$ and $h'_0 =$* II *defined on a neighbourhood M of $\{0\}\times\Sigma$ in $\mathbb{R}\times\Sigma$, such that (M,g) is globally hyperbolic with Cauchy hypersurface Σ.*

Proof We proceed in three steps:

Step 1: Let $p \in \Sigma$. We fix coordinates charts of Σ around p and apply the Cauchy-Kowalevski Theorem to the evolution equation (37) written in these coordinates. On each coordinate patch, this will give a unique real analytic solution h_t defined on a neighbourhood of U_p in $\mathbb{R}\times\Sigma$. By Lemma 1 this neighbourhood U_p can be chosen to be globally hyperbolic. By the uniqueness, the solutions on such globally hyperbolic neighbourhoods for different p and q coincide on their overlaps $U_p \cap U_q$ and hence define a unique real analytic solution h_t on neighbourhood $M := \cup_{p\in\Sigma} U_p$ of $\{0\}\times\Sigma$ in $\mathbb{R}\times\Sigma$.

Step 2: We show that (M,g) is globally hyperbolic. Let $\gamma = (\gamma_0, \gamma_\Sigma) : (a,b) \to M$ be an inextendible timelike geodesic. Then γ intersect one of the globally hyperbolic neighbourhoods from Step 1 and hence intersects Σ, i.e., there is a $c \in (a,b)$ such that $\gamma_0(c) = 0$. Since γ is timelike, it is $\gamma'_0(s) \neq 0$ for all $s \in (a,b)$ which implies that γ_0 cannot return to 0, and hence $\gamma(c)$ is the only intersection of γ with Σ.

Step 3: It remains to show that the resulting Lorentzian manifold (M,g) is actually Ricci-flat. The evolution equation ensures that $Ric^g|_{T\Sigma_t\times T\Sigma_t} = 0$. We define a function and a 1-form on M by

$$f = Ric^g(T,T) = -scal^g, \qquad \omega = Ric^g(T,.),$$

and show that they are zero. The key is the second Bianchi-identity and its consequence (1). Using $Ric^g|_{T\Sigma_t \times T\Sigma_t} = 0$, it provides us with evolution equations for f and ω,

$$f' = 2\mathrm{tr}(W)f + 2\mathrm{div}^{h_t}(\omega) + 4\omega(\mathrm{grad}^{h_t}(\log(\lambda)))$$
$$\omega' = \mathrm{tr}(W)\omega + \tfrac{1}{2}df.$$

Obviously, the right-hand-side is $\mathbb{R}$-linear in f and ω. To these equations we can again apply the Cauchy-Kowalevski Theorem. The constraint equations along Σ imply that $f|_\Sigma = 0$ and $\omega|_\Sigma = 0$, so that the unique solutions for f and ω vanish on all of M. This concludes the proof. □

This result goes back to Darmois [34], see also [64]. It is independent of the signature of g and in particular holds for Riemannian metrics $g = \lambda^2 dt^2 + h_t$, see [3, 56].

6.4 The Vacuum Einstein Equations as Symmetric Hyperbolic System

The generalisation of Theorem 20 to the smooth setting goes back to Y. Choquet-Bruhat [39]. In order to apply the theory of symmetric hyperbolic PDEs to the vacuum Einstein equation, one approach is to break the diffeomorphism invariance. This can be done done by fixing certain coordinates, Choquet-Bruhat used *harmonic coordinates*, [39], another approach is to fix a *background metric*. This is known as *hyperbolic reduction*, and goes back to Friedrich and Renadall [42], see also [70].

For a Lorentzian manifold (M, g), recall from Sect. 6.2 that the term that prevents Eq. (33) from being symmetric hyperbolic when written in coordinates was $\partial_{(\mu}\Gamma_{\nu)}$, see Eq. (43). This term cannot be expressed invariantly, so just subtracting this term from the Ricci tensor would not give an invariant PDE, it suggest however a modification of Ric^g by the symmetric derivative of a 1-form η,

$$Ric^g + \mathrm{Sym}(\nabla^g \eta),$$

where the symbol Sym denotes the symmetrisation of a bilinear form $\mathrm{Sym}(B)(X, Y) = \frac{1}{2}(B(X, Y) + B(Y, X))$. The idea to find η in [42], known as *hyperbolic reduction*, is to fix a background metric $\widetilde{g}$ with Levi-Civita connection $\widetilde{\nabla}$ and define η by taking its trace with respect its two covariant entries and dualising with the metric g give a 1-form

$$\eta^{g,\widetilde{g}}(X) := \sum_{\mu=0}^{n} \epsilon_\mu g(C(E_\mu, E_\mu), X), \tag{53}$$

for $X \in TM$, E_μ an orthonormal basis for g, and where C is the difference tensor of ∇^g and $\widetilde{\nabla}$,

$$C(X,Y) = \widetilde{\nabla}_X Y - \nabla^g_X Y.$$

Then we define

$$\widetilde{Ric}^g := Ric^g + \mathrm{Sym}(\nabla^g \eta^{g,\widetilde{g}}), \tag{54}$$

In the following, we use again Einstein's summation convention and the convention that Greek indices run from $0,\dots,n$ and Latin indices from $1,\dots,n$.

Lemma 8 *Let $(M,\widetilde{g})$ be a Lorentzian manifold and g Lorentzian metric on M satisfying the equation*

$$\widetilde{Ric}^g = 0, \tag{55}$$

where $\widetilde{Ric}^g$ is defined in (54). Then in local coordinates x^μ, with x^0 a timelike coordinate, the metric coefficients $g_{\mu\nu}$ and its first derivatives $K_{\mu\nu} := \partial_0 g_{\mu\nu}$ and $L_{\mu\nu,i} := \partial_i g_{\mu\nu}$ satisfy a symmetric hyperbolic system of the following form

$$\begin{aligned} \partial_0 g_{\mu\nu} &= K_{\mu\nu}, \\ g^{00}\partial_0 K_{\mu\nu} + 2g^{0i}\partial_i K_{\mu\nu} + g^{ij}\partial_i L_{\mu\nu,j} &= F_{\mu\nu}, \\ -g_i{}^j \partial_0 L_{\mu\nu,j} + g_i{}^j \partial_j K_{\mu\nu} &= 0, \end{aligned} \tag{56}$$

with suitable functions $F_{\mu\nu} = F_{\mu\nu}(x^\alpha, g_{\alpha\beta}, K_{\alpha\beta}, L_{\alpha\beta,k})$ depending on the coordinates (x^α), the metric coefficients and ist first derivatives.

Proof In local coordinates (x^μ) the term $\mathrm{Sym}(\nabla^g\eta)$ in (54) is

$$\nabla^g_{(\mu}\eta_{\nu)} = \tfrac{1}{2}g^{\alpha\beta}\left(\partial_\mu\partial_\nu g_{\alpha\beta} - \partial_\mu\partial_\alpha g_{\beta\nu} - \partial_\mu\partial_\beta g_{\alpha\nu}\right) + LOTs,$$

and hence, because of Eq. (44), the functions

$$f_{\mu\nu} := \partial_{(\mu}\Gamma_{\nu)} + \nabla^g_{(\mu}\eta_{\nu)}$$

are of lower order in $g_{\alpha\beta}$, i.e., $f_{\mu\nu} = f_{\mu\nu}(x^\alpha, g_{\alpha\beta}, \partial_\gamma g_{\alpha\beta})$. Hence, because of Eq. (43), we get

$$(\widetilde{Ric}^g)_{\mu\nu} = -\tfrac{1}{2}g^{\alpha\beta}\partial_\alpha\partial_\beta g_{\mu\nu} + F_{\mu\nu}, \tag{57}$$

where $F_{\mu\nu} = f_{\mu\nu} + G_{\mu\nu}$ and the $G_{\mu\nu}$'s are of lower order in $g_{\alpha\beta}$ and defined in Eq. (43).

Then Eq. (55) and the definition of the $K_{\mu\nu}$'s and $L_{\mu\nu,i}$'s imply the system (56). It remains to show that (56) is symmetric hyperbolic.

To this end, on the appropriate vector spaces we define linear maps $A^0 = A^0(x^\alpha, g_{\alpha\beta}, K_{\alpha\beta}, L_{\alpha\beta,k})$ and $A^i = A^i(x^\alpha, g_{\alpha\beta}, K_{\alpha\beta}, L_{\alpha\beta,k})$, depending on the coordinates, the metric coefficients and its derivatives, as

$$A^0 \begin{pmatrix} u_{\mu\nu} \\ v_{\mu\nu} \\ w_{\mu\nu,i} \end{pmatrix} := \begin{pmatrix} u_{\mu\nu} \\ g^{00} v_{\mu\nu} \\ -g_i^{\ j} w_{\mu\nu,j} \end{pmatrix}, \qquad A^i \begin{pmatrix} u_{\mu\nu} \\ v_{\mu\nu} \\ w_{\mu\nu,j} \end{pmatrix} := \begin{pmatrix} 0 \\ 2g^{0i} v_{\mu\nu} + g^{ik} w_{\mu\nu,k} \\ g^i_{\ j} v_{\mu\nu} \end{pmatrix}.$$

Writing these linear maps schematically as matrices shows that they are symmetric,

$$A^0 = \begin{pmatrix} 1 & 0 & 0 \\ 0 & g^{00} & 0 \\ 0 & 0 & -g_i^{\ j} \end{pmatrix}, \qquad A^i = \begin{pmatrix} 0 & 0 & 0 \\ 0 & 2g^{0i} & g^{ij} \\ 0 & g^{ij} & 0 \end{pmatrix}.$$

Moreover, since g is Lorentzian metric and the x^0 coordinate is timelike, the matrix $-g_{ij}$ is negative definite and hence A^0 is positive definite. It is easily checked that the system (56) can be written as

$$A^0 \begin{pmatrix} \partial_0 g_{\mu\nu} \\ \partial_0 K_{\mu\nu} \\ \partial_0 L_{\mu\nu,j} \end{pmatrix} + A^i \begin{pmatrix} \partial_i g_{\mu\nu} \\ \partial_i K_{\mu\nu} \\ \partial_i L_{\mu\nu,j} \end{pmatrix} = b$$

as in (42) with

$$b = b(x^\alpha, g_{\alpha\beta}, K_{\alpha\beta}, L_{\alpha\beta,k}) = \begin{pmatrix} K_{\mu\nu} \\ F_{\mu\nu}(x^\alpha, g_{\alpha\beta}, K_{\alpha\beta}, L_{\alpha\beta,k}) \\ 0 \end{pmatrix},$$

where $F_{\mu\nu}$ is determined by the LOTs in Eq. (57). □

A few remarks are in place. First, it is clear that the system (56) consist of less equations than the original system (57), which was equivalent to Eq. (55) and hence is weaker. We will however see later that Eq. (56) together with the appropriate initial conditions implies Eq. (55).

Secondly, clearly a solution g to equation (55) is not necessarily Ricci-flat. The key observation then is that if g satisfies condition (55) then the 1-form η satisfies a *linear wave equation*.

Lemma 9 *Let (M, g) be a Lorentzian manifold with Ricci tensor Ric^g and η be a solution to equation*

$$Ric^g + \mathrm{Sym}(\nabla^g \eta) = 0. \tag{58}$$

The η satisfies the wave equation

$$\Delta_g \eta - Ric^g(\eta^\sharp, .) = 0,$$

where Δ_g denotes the Bochner-Laplacian on 1-forms, $\Delta_g \eta = \mathrm{tr}^g_{(1,2)}(\nabla\nabla\eta)$, or in abstract index notation $(\Delta_g \eta)_\mu = \nabla_\nu \nabla^\nu \eta_\mu$.

Proof Taking the trace of Eq. (58) gives

$$0 = scal^g + \mathrm{div}^g(\eta).$$

Taking the divergence of (58), the identity (1) and the definition of the curvature yield the simple computation

$$\begin{aligned} 0 &= -\nabla_\mu scal^g + \nabla_\nu \nabla^\nu \eta_\mu + \nabla_\nu \nabla^\mu \eta^\nu \\ &= -\nabla_\mu \nabla_\nu \eta^\nu + \nabla_\nu \nabla^\nu \eta_\mu + \nabla_\nu \nabla^\mu \eta^\nu \\ &= \nabla_\nu \nabla^\nu \eta_\mu + R_{\nu\mu}{}^{\nu}{}_{\alpha} \eta^\alpha \\ &= \Delta_g \eta_\mu - Ric^g_{\mu\alpha} \eta^\alpha, \end{aligned}$$

which proves the lemma. □

By Theorem 3, this implies that η is uniquely determined by its values and derivatives on an initial Cauchy hypersurface. This means that one has to choose the initial values such that $\eta|_\Sigma = 0$ and $\nabla\eta|_\Sigma = 0$ for Σ. The following will facilitate this:

Lemma 10 *Let (M, g) be a Lorentzian manifold with Ricci tensor Ric^g and η be a solution to Eq. (58). Moreover, let (Σ, h) be a spacelike hypersurface satisfying the constraint equations (39) and (40). If $\eta|_\Sigma = 0$, then also $\nabla^g \eta|_\Sigma = 0$.*

Proof Because of $\eta|_\Sigma = 0$, we also have that $\nabla^g_X \eta|_\Sigma = 0$ for all $X \in T\Sigma$. For the other terms, the constraint conditions come into play: by Lemma 6, the constraint condition (39) gives that

$$0 = Ric^g(\partial_t, X) = -\tfrac{1}{2}(\nabla^g_{\partial_t} \eta)(X),$$

where we also use Eq. (55). Equation (55) and what we know about η so far also give

$$Ric^g(T, T)|_\Sigma = -(\nabla^g_T \eta)(T)|_\Sigma = \mathrm{div}^g(\eta)|_\Sigma = -scal^g|_\Sigma.$$

But then Eq. (38) and the constraint condition (40) show that $scal^g$ and hence $\nabla^g \eta$ vanish along Σ. □

We will now specify the correct initial conditions that enable us to put Lemmas 8–10 into action.

Theorem 21 *Let (Σ, h) be a Riemannian manifold equipped with a symmetric bilinear form* II, *a smooth function $f \in C^\infty(\Sigma)$ and a smooth* 1*-form σ. Moreover, let λ be a smooth positive function on $\mathbb{R} \times \Sigma$. On $\mathbb{R} \times \Sigma$ fix the background metric*

$$\widetilde{g} = -\lambda^2 dt^2 + h. \tag{59}$$

Then on a neighbourhood M of $\{0\} \times \Sigma$ in $\mathbb{R} \times \Sigma$ there is a unique Lorentzian metric g that satisfies Eq. (55) and the initial conditions

$$g|_\Sigma = \widetilde{g}|_\Sigma, \quad g'|_{T\Sigma \times T\Sigma} = -2\lambda|_\Sigma \mathrm{II}, \qquad g'(\partial_t, \partial_t)|_\Sigma = f, \quad g'(\partial_t, .)|_\Sigma = \sigma,$$

and such that (M, g) contains (Σ, h) as Cauchy hypersurface.

Proof The proof proceeds similarly to the proof of Theorem 20. At $p \in \Sigma$ we fix coordinates x^i on an open set W in Σ and denote by $x^0 := t$ the canonical t-coordinate on $\mathbb{R} \times \Sigma$. Since the system (56) turned out to be symmetric hyperbolic, it has a unique solution $(g_{\mu\nu}, K_{\mu\nu}, L_{\mu\nu,i})$ on a neighbourhood U of W in $\mathbb{R} \times W$ satisfying the initial conditions

$$g_{\mu\nu}|_W = \widetilde{g}_{\mu\nu}|_W, \quad K_{ij}|_W = -2\lambda \mathrm{II}_{ij}, \quad K_{00}|_W = f, \quad K_{0i}|_W = \sigma_i,$$

as well as $L_{\mu\nu,k}|_W = \partial_k \widetilde{g}_{\mu\nu}|_W$, i.e.,

$$L_{ij,k}|_W = \partial_k h_{ij}, \quad L_{i0,k}|_W = 0, \quad L_{00,i}|_W = \partial_i \lambda|_W. \tag{60}$$

This first equation in (56) is just that $K_{\mu\nu} = \partial_0 g_{\mu\mu}$, which turns the third into

$$0 = \partial_0 \left(L_{\mu\nu,i} - \partial_i g_{\mu\nu} \right).$$

This, together with the first initial condition in (60) implies that $L_{\mu\nu,i} = \partial_i g_{\mu\nu}$ on all of U. With this, the second equation in the system (56) just becomes the the second order equation (57) and hence implies Eq. (55). Hence for each $p \in \Sigma$ we obtain a Lorentzian metric on an open neighbourhood U_p in $\mathbb{R} \times W$ satisfying Eq. (55). As before, we can chose U_p to be globally hyperbolic, but in addition we chose it so small that $U_p \cap \Sigma_t$ is spacelike, where $\Sigma_t = \{t\} \times \Sigma$. This requirement will be needed in the next step to ensure that the t-component of a timelike geodesic is nonzero.

Having such U_p for each $p \in \Sigma$, we define now $M := \cup_{p \in \Sigma} U_p$. As before, the uniqueness of the solution implies that two solutions on different U_p coincide on overlaps and hence constitute a solution on a neighbourhood $M = \cup_{p \in M} U_p$ of Σ in $\mathbb{R} \times \Sigma$. As before in Step 2 in the proof of Theorem 20 one can now show that M is globally hyperbolic, using that the U_p are globally hyperbolic and the additional requirement that $\Sigma_t \cap U_p$ is spacelike. The latter ensures that each curve $\gamma = (\gamma_0, \gamma_\Sigma)$ satisfies $g(\gamma'_\Sigma(s), \gamma'_\Sigma(s) \geq 0$, which implies for a timelike curve that

$\gamma_0'(s) \neq 0$ for all s. Hence the argument in Step 2 in the proof of Theorem 20 can be applied to show that (M, g) is globally hyperbolic. □

In order to specify the remaining initial conditions f and σ in Theorem 21 in a way that the 1-form $\eta^{g,\widetilde{g}}$ vanishes, we will make use of the following observation:

Lemma 11 *If $\widetilde{g}$ is the background metric (59) on $\mathbb{R} \times \Sigma$ and g another Lorentzian metric on a neighbourhood of Σ in $\mathbb{R} \times \Sigma$ such that $g|_\Sigma = \widetilde{g}|_\Sigma$, then $\eta = \eta^{g,\widetilde{g}}$ satisfies*

$$\begin{aligned} \eta(\partial_t)|_\Sigma &= \tfrac{1}{2\lambda^2} g'(\partial_t, \partial_t)|_\Sigma + \tfrac{1}{2}\mathrm{tr}^h(g'|_\Sigma) + \tfrac{\lambda'}{\lambda}|_\Sigma, \\ \eta|_{T\Sigma} &= \tfrac{1}{\lambda^2} g'(\partial_t, .)|_{T\Sigma}. \end{aligned}$$

Proof We use coordinates $x^0 = t$ and x^i coordinates on Σ. Then, because of the initial conditions $g|_\Sigma = \widetilde{g}|_\Sigma$, we have *along Σ*

$$\widetilde{\Gamma}^i_{kl} - \Gamma^i_{kl} = 0, \quad \widetilde{\Gamma}^0_{kl} - \Gamma^0_{kl} = -\tfrac{g'_{kl}}{2\lambda^2}, \quad \widetilde{\Gamma}^0_{00} - \Gamma^0_{00} = \tfrac{\lambda'}{\lambda} + \tfrac{g'_{00}}{2\lambda^2}, \quad \widetilde{\Gamma}^k_{00} - \Gamma^k_{00} = -h^{kj} g'_{0j}.$$

Hence, still along Σ we get

$$\eta_0|_\Sigma = g^{\mu\nu} g_{0\alpha} \left(\widetilde{\Gamma}^\alpha_{\mu\nu} - \Gamma^\alpha_{\mu\nu}\right)|_\Sigma = \widetilde{\Gamma}^0_{00} - \Gamma^0_{00} - \lambda^2 h^{kl} \left(\widetilde{\Gamma}^0_{kl} - \Gamma^0_{kl}\right)|_\Sigma,$$

which proves the first equation, and

$$\eta_i|_\Sigma = g^{\mu\nu} g_{i\alpha} \left(\widetilde{\Gamma}^\alpha_{\mu\nu} - \Gamma^\alpha_{\mu\nu}\right)|_\Sigma = g^{\mu\nu} h_{ik} \left(\widetilde{\Gamma}^k_{\mu\nu} - \Gamma^k_{\mu\nu}\right)|_\Sigma = \lambda^{-2} g'_{0i}|_\Sigma,$$

proving the second one. □

Now have all we need in order to prove:

Theorem 22 *Let (Σ, h) be a Riemannian manifold satisfying the constraint conditions (40) and (39) for a symmetric bilinear form II and let λ be a smooth positive function on $\mathbb{R} \times \Sigma$ and $\widetilde{g}$ the associated background metric (59). Then on a neighbourhood M of $\{0\} \times \Sigma$ in $\mathbb{R} \times \Sigma$ there is a unique Ricci-flat Lorentzian metric g with $\eta^{g,\widetilde{g}} = 0$ and satisfying the initial conditions*

$$g|_\Sigma = \widetilde{g}|_\Sigma, \qquad g'|_\Sigma = -2\lambda \mathrm{II} + 2\lambda^2 \left(\lambda \mathrm{tr}^h(\mathrm{II}) - \lambda'\right) dt^2.$$

Moreover (M, g) contains (Σ, h) as Cauchy hypersurface.

Proof The initial conditions are as in Theorem 21 with

$$f := 2\lambda^2 \left(\lambda \mathrm{tr}^h(\mathrm{II}) - \lambda'\right), \qquad \sigma := 0.$$

By Theorem 21 we get a unique globally hyperbolic Lorentzian manifold (M, g) with $\widetilde{Ric}^g = 0$. By Lemma 11 the initial conditions ensure that $\eta^{g,\widetilde{g}}|_\Sigma = 0$. By Lemma 10 this also implies that $\nabla^g \eta^{g,\widetilde{g}}|_\Sigma = 0$. Now we can apply Lemma 9 and Theorem 3 to conclude that $\eta^{g,\widetilde{g}} = 0$ on all of M and hence that g is Ricci flat. □

Remark 5 In contrast to the result in Theorem 20, the Ricci-flat metric found in Theorem 22 is not necessarily of the form $-\lambda^2 dt^2 + h_t$. We can however arrange with a simple trick that the hypersurfaces Σ_t remain orthogonal to ∂_t, see [62, Section 5]:

If t denotes the global t-coordinate on $M \subset \mathbb{R} \times \Sigma$, its gradient with respect to g is by construction timelike and the leaves of the integrable distribution $\mathrm{grad}^g(t)^\perp$ are given by $\Sigma_t = \{t\} \times \Sigma$. Then the vector field $F = (dt(\mathrm{grad}^g(t))^{-1}\,\mathrm{grad}^g(t)$ satisfies $dt(F) = 1$ and its flow ϕ_s sends level sets Σ_t to Σ_{t+s}. Indeed, the derivative of the function $f(s) := t(\phi_s(p))$ satisfies

$$\tfrac{df}{ds} = dt|_{\phi_s(p)}(F) = 1,$$

and hence $f(s) = t(\phi_s(p)) = s + t(p)$. The flow ϕ_t defines a diffeomorphism ψ on a neighbourhood of Σ in M by $\psi(t, p) = \phi_t t(p)$ that satisfies $d\psi(\partial_t) = F$. Hence the pulled back metric is of the form

$$\psi^* g = -\widetilde{\lambda}^2 dt^2 + h_t,$$

with a smooth function $\widetilde{\lambda}$. Along Σ it satisfies $\widetilde{\lambda}|_\Sigma = g(\mathrm{grad}^g(t), \mathrm{grad}^g(t))^{-1/2}|_\Sigma = \lambda|_\lambda$, i.e., along Σ, $\widetilde{\lambda}$ coincides with the prescribed function λ. Off Σ, it might however differ from λ.

7 Cauchy Problems for Lorentzian Special Holonomy

In this section we will see that the constraint conditions in Sect. 5 for a Riemannian manifold (Σ, h) are not only necessary but also sufficient for (Σ, h) being a Cauchy hypersurface in a Lorentzian manifold with parallel lightlike vector field (spinor field). Following the strategy we have described in the previous section for the vacuum Einstein equations, we will achieve this by studying certain Cauchy problems for PDEs that are induced by the existence of parallel lightlike vector fields or spinor fields. This will involve finding the right evolution equations for a time-dependent family of Riemannian metrics as well as ensure that these equations have a solution. As before we will study this in the analytic setting where we derive evolution equations [16] and in the smooth setting where we study a symmetric hyperbolic PDE system [62].

7.1 Evolution Equations for a Parallel Lightlike Vector Field in the Analytic Setting

As in Sect. 6.3 in this section we again make the ansatz to construct a Lorentzian metric $g = -\lambda^2 dt^2 + h_t$ of the form (45) on an open neighbourhood M in $\mathbb{R} \times \Sigma$ of a Riemannian manifolds (Σ, h) satisfying the constraint condition (25) and (24).

We assume that $M = \mathbb{R} \times \Sigma$ and $g = -\lambda^2 dt^2 + h_t$. If we write a lightlike vector field V as $V = u_t T - U_t$ with t-dependent function and vector field on Σ, then the constraints for each t,

$$\nabla^{h_t} U_t + u_t W_t = 0, \qquad du_t + \mathrm{II}_t(U_t, \cdot) = 0$$

are equivalent to $\nabla^g V|_{\Sigma_t} = 0$, whereas $\nabla^g_{\partial_t} V = 0$ is equivalent to

$$U_t' = \lambda W(U) + u_t\, \mathrm{grad}^{h_t}(\lambda), \qquad u' = d\lambda(U_t), \tag{61}$$

where the prime denotes the t-derivative, i.e., $U' = [\partial_t, U]$. Equation (61), considered as an equation for U, u and g however cannot be brought into Cauchy-Kowalevski form. Instead we will find a second order equation for U, u and g resulting from $V \lrcorner R^g = 0$ and show that we can apply the Cauchy-Kowalevski Theorem to it. The key to find the evolution equations is the following observation:

Lemma 12 *A vector field V is parallel on (M, g) if and only if the following equations are satisifed*

$$R^g(\partial_t, X)V = 0, \quad \textit{for all } X \in \partial_t^\perp = T\Sigma_t, \tag{62}$$

$$\nabla^g_{\partial_t} \nabla^g_{\partial_t} V = 0, \tag{63}$$

$$\nabla^g_X V|_\Sigma = 0, \quad \textit{for all } X \in T\Sigma \tag{64}$$

$$\nabla^g_{\partial_t} V|_\Sigma = 0. \tag{65}$$

Proof If V is parallel, all the conditions follow immediately. On the other hand, Eq. (63) shows that the vector field $\nabla^g_{\partial_t} V$ is parallel transported along the curves $t \mapsto (t, x)$. Hence, with the initial condition (65), we have $\nabla^g_{\partial_t} V = 0$ everywhere on M. Using this, Eq. (62) for vector fields $X \in \Gamma(\partial_t^\perp)$ with $[\partial_t, X] = 0$ gives that $0 = R^g(\partial_t, X)V = \nabla^g_{\partial_t} \nabla^g_X V$, which shows that $\nabla^g_X V$ is parallel transported along all $t \mapsto (t, x)$. Since we have assumed that $\nabla^g_X V = 0$ along the initial manifold Σ, it also shows that V is parallel on M. □

In these equations, (64) just is the constraint condition (61) for $t = 0$, Eq. (65) gives initial conditions for $U|_{t=0}$, and Eqs. (62) and (63) provide us with evolution equations for h_t, U_t and u_t. For this we have to rewrite (62) using the Codazzi-Mainardi equations and obtain that Eq. (62) is equivalent to

$$u_t \mathrm{II}_t'(X, Y) = \lambda (d^{h_t} \mathrm{II}_t)(U_t, Y, X) - u_t \lambda \mathrm{II}_t(X, W_t(Y)) - u_t \mathrm{Hess}^{h_t}(\lambda)(X, Y), \tag{66}$$

for all $X, Y \in \partial_t^\perp$ and where $d^{h_t}\mathrm{II}_t$ is defined by

$$(d^{h_t}\mathrm{II}_t)(X, Y, Z) = \tfrac{1}{2}\left((\nabla_X^{h_t}\mathrm{II}_t)(Y, Z) - (\nabla_Y^{h_t}\mathrm{II}_t)(X, Z)\right).$$

When spelling out the II'-terms, Eq. (66) becomes a second order evolution equation for the metric h_t which is of Cauchy-Kowalevski form. Moreover, the right-hand-side does not contain any time derivatives of U_t or u_t. There is however an obvious problem with this equation: the term $(d^{\nabla_t}\mathrm{II}_t)(U_t, Y, X)$ is not necessarily symmetric in X and Y. In fact, its skew-symmetrisation is equal to $d^{h_t}(X, Y, U_t)$ which in general is not zero. Hence, when considered as an evolution equation for g, it might yield a solution that is not symmetric. We can resolve this problem under the assumption that all data are real analytic.

Lemma 13 *Let (M, g) be real analytic and let V be an analytic lightlike vector field on M. Then $\nabla^g V = 0$, if and only if Eqs. (63)–(65) together with*

$$R^g(X, V, V, Y) = 0, \quad \text{for all } X, Y \in \partial_t^\perp, \tag{67}$$

are satisfied.

Proof (Sketch of the Proof) Having Eqs. (63)–(65), but not Eq. (62) it remains to show that $\nabla_X^g V = 0$ on all of M. The idea is to use Eq. (67) to derive a linear system of the form

$$\nabla_{\partial_t}^g \begin{pmatrix} A \\ B \\ C \end{pmatrix} = Q \begin{pmatrix} A \\ B \\ C \end{pmatrix},$$

for $A, B \in (\partial_t^\perp)^* \otimes TM$ and $C \in \Lambda^2(\partial_t^\perp)^* \otimes TM$ defined by

$$A(X) = \nabla_X^g V, \quad B(X) = R^g(T, X)V, \quad C(X, Y) = R^g(X, Y)V.$$

and where Q is a linear operator on an appropriate vector space. Moreover, the initial conditions (64) and (65) imply that A, B, and C vanish along Σ. Hence, being subject to the linear system above, they vanish everywhere. □

Now it is a matter of determining the explicit forms of the evolution equations. For Eq. (67) one uses again the Codazzi-Mainardi equations and obtains an equation similar to (66). Equation (63) on the other hand, when written in terms of U_t and u_t is equivalent to the equations

$$\begin{aligned} U_t'' &= \lambda\left([\partial_t, W_t(U_t)] + W_t(U_t') - \lambda W_t^2(U_t)\right) + \lambda_t' W_t(U) \\ &\qquad + u_t\left([\partial_t, \mathrm{grad}^{h_t}\lambda] - \lambda W_t(\mathrm{grad}^{h_t}\lambda)\right) + \left(2u_t' - d\lambda(U_t)\right)\mathrm{grad}^{h_t}\lambda \\ u_t'' &= h_t([\partial_t, \mathrm{grad}^{h_t}\lambda], U_t) + 2d\lambda(U_t') - 3\lambda d\lambda(W_t(U_t)) - u_t\|\mathrm{grad}^{h_t}\lambda)\|_{h_t}^2. \end{aligned}$$

The crucial observation is that the right-hand-sides do not contain any second t-derivatives, so that we can apply the Cauchy-Kowalevski Theorem.

Theorem 23 *Let (Σ, h, W, U) be an analytic Riemannian manifold together with a field of h-symmetric, analytic endomorphisms W, with corresponding symmetric bilinear form $\mathrm{II} := h(W.,.)$, and an analytic vector field U satisfying the following constraint equation*

$$\nabla^h U + uW = 0, \tag{68}$$

where $u^2 = h(U, U)$. Then, for any positive analytic function λ on $\mathbb{R} \times \Sigma$ there exists an open neighbourhood M of $\{0\} \times \Sigma$ in $\mathbb{R} \times \Sigma$ and an unique analytic Lorentzian metric

$$g = -\lambda^2 dt^2 + h_t$$

on M which admits an analytic lightlike parallel vector field $V = \frac{u_t}{\lambda}\partial_t - U_t$ and satisfies the initial conditions

$$\begin{aligned} h_0 &= h, & U_0 &= U, & u_0 &= u, \\ h_0' &= -2\lambda_0 \mathrm{II}, & U_0' &= u\,\mathrm{grad}^h(\lambda_0) + \lambda_0 W(U), & u_0' &= d\lambda_0(U). \end{aligned} \tag{69}$$

Moreover, M can be chosen such that $\Sigma \subset M$ is a spacelike Cauchy hypersurface. The t-dependent families of metrics h_t, vector fields U_t and smooth functions u_t on Σ are solutions to the following PDE system

$$\begin{aligned} h_t''(X,Y) = {} & \frac{\lambda^2}{u_t}\left(d^{h_t}\Big(\frac{h_t'}{\lambda}\Big)(U_t, X, Y) + d^{h_t}\Big(\frac{h_t'}{\lambda}\Big)(U_t, Y, X)\right) + \frac{1}{2}h_t'(X, (h_t')^\sharp(Y)) \\ & + (\log\lambda)' h_t'(X,Y) + 2\lambda \mathrm{Hess}^{h_t}(\lambda)(X,Y) + 2\frac{\lambda^2}{u_t^2} R^{h_t}(X, U_t, U_t, Y) \\ & + \frac{1}{2u_t^2}\Big(h_t'(X,Y)h_t'(U_t,U_t) - h_t'(X,U_t)h_t'(Y,U_t)\Big), \end{aligned}$$

$$\begin{aligned} h_t(U_t'', X) = {} & -\frac{\lambda^2}{2u_t} d^{h_t}\Big(\frac{h_t'}{\lambda}\Big)(U_t, X, U_t) - \frac{1}{2}(\log\lambda)' h_t'(U_t, X) - \lambda \mathrm{Hess}^{h_t}(\lambda)(U_t, X) \\ & - h_t'(U_t', X) + u_t h_t([\partial_t, \mathrm{grad}^{h_t}\lambda], X) + \frac{u_t}{2} h_t'(\mathrm{grad}^{h_t}\lambda, X) \\ & + \big(2u_t' - d\lambda(U_t)\big)\, d\lambda(X), \end{aligned}$$

$$u_t'' = h_t([\partial_t, \mathrm{grad}^{h_t}\lambda], U_t) + 2d\lambda(U_t') + \frac{3}{2}h_t'(\mathrm{grad}^{h_t}(\lambda).U_t) - u_t \|\mathrm{grad}^{h_t}\lambda\|_{h_t}^2.$$

The proof is analogous to the proof of Theorem 20 for the Einstein equations and follows from using the Cauchy-Kowalevsky theorem locally and then obtaining a globally hyperbolic manifold by the uniqueness of the solutions.

Finally, we give an example which was first given in [13] and for which the above system can be solved explicitly.

Example 10 Let (M, h, W, U, u) be a Riemannian manifold with a symmetric endomorphism field W, a vector field U and a function u on M satisfying the constraint equations

$$\nabla^h U = -uW, \qquad h(U, U) = u^2 > 0.$$

If, in addition, W is a Codazzi tensor, i.e., $d^{\nabla^h} W = 0$, and $\lambda = 1$, then the following are solutions to the PDE system in Theorem 23

$$\begin{aligned} h_t &:= h - 2th(W(\cdot), \cdot) + t^2 h(W^2(\cdot), \cdot) = h((1 - tW)^2(\cdot), \cdot), \\ U(t, x) &:= \frac{1}{(1 - tW_x)} U(x) = \sum_{k=0}^{\infty} W_x^k (U(x)) t^k, \\ u(t, x) &:= u(x). \end{aligned}$$

They are defined on

$$\mathcal{M} := \{(t, x) \in \mathbb{R} \times \Sigma \mid t\|W_x\|_{h_x} < 1\}.$$

7.2 The Cauchy Problem for a Parallel Lightlike Vector Field as a Symmetric Hyperbolic System

Now we want to present recent results in [62] that generalise Theorem 23 to the smooth setting. We want to take the same approach as for the Cauchy problem for the Einstein equations in Sect. 6.4, and moreover we would like to partially reduce the Cauchy problem for parallel lightlike vector fields to the vacuum Einstein equations. But, whereas the integrability conditions for parallel lightlike spinor fields in Proposition 4 are formulated in terms of the Ricci tensor and lead to obvious evolution equations for the metric in the Cauchy problem for parallel lightlike spinor fields (in [65], see also Remark 7 below), the existence of a parallel lightlike vector field yields hardly any nontrivial information about the Ricci tensor. Thus, it is not obvious at all that the methods that work for the Cauchy problem for the Einstein equations or a parallel lightlike spinor field also work for a parallel lightlike vector field and that Theorem 23 can be generalised to the smooth setting. The idea here is to simply introduce the Ricci tensor as new unknown, i.e. set $Ric^g = Z$. We consider this as an evolution equation for the metric g and close the system by deriving a first order equation for Z by further differentiation.

To find the PDE that will form the correct quasilinear first order hyperbolic system, we assume that (Σ, h) is a Riemannian manifold and that M is an open neighbourhood of Σ in $\mathbb{R} \times \Sigma$ with a Lorentzian metric g, a function $\lambda = \sqrt{-g(\partial_t, \partial_t)}$, and such that along Σ, which we identify within M with $\Sigma_0 = \{0\} \times \Sigma$, it is

$$g|_\Sigma = -\lambda^2|_\Sigma dt^2 + h,$$

i.e., such that the initial hypersurface Σ is g-orthogonal to ∂_t. In contrast to Sect. 7.1, however we will not assume that $\Sigma_t = \{t\} \times \Sigma$ is g-orthogonal to ∂_t away from $\Sigma_0 = \Sigma$. We denote by $\pi : M \ni (t, p) \to p \in \Sigma$ the projection onto the second component and by $\mathrm{pr}_{T\Sigma} = d\pi$ its differential, i.e., the g-independent projection

$$\mathrm{pr}_{T\Sigma} : TM = \mathbb{R}\partial_t \oplus T\Sigma \to T\Sigma$$

onto the second factor.

Now we assume that (M, g) admits a parallel lightlike vector field V. We set $U := -\mathrm{pr}_{T\Sigma}(V) \in \Gamma(\pi^* T\Sigma)$. Hence, using the timelike unit vector field $T = \frac{1}{\lambda}\partial_t$, V decomposes into

$$V = vT - U, \tag{70}$$

with $v \in C^\infty(M)$. Note that $V \neq 0$ and $g(V, V) = 0$ imply that v, U and hence $u := \sqrt{g(U, U)}$ have no zeros. Along the initial hypersurface Σ we have that $g(T, U)_{|\Sigma_0} = 0$, as T was assumed to be the unit normal vector field along Σ, and hence that $v|_\Sigma = u|_\Sigma$. Moreover we have seen that in this situation $(\Sigma, h, U|_\Sigma)$ satisfies the constraints (25) and (24).

We will now derive and analyse several PDEs that follow from $\nabla^g V = 0$ and which will constitute a symmetric hyperbolic system. First note that the 1-form $V^\flat$ satisfies the PDE

$$(d + \mathrm{div}^g)\omega = 0, \tag{71}$$

Here $d + \mathrm{div}^g : \Omega^*(M) \to \Omega^*(M)$ is the de Rham operator on the algebra of forms. It is also given by

$$d + \mathrm{div}^g = c \circ \nabla,$$

where $c : TM \otimes \Lambda^* M \to \Lambda^* M$ denotes Clifford multiplication by forms, i.e.,

$$c(X) \cdot \omega = X^\flat \wedge \omega - \iota_X \omega, \text{ for all } X \in TM, \tag{72}$$

where $\iota_X\omega = \omega(X,\ldots)$ denotes the interior product. The advantage of this equation is that $d + \mathrm{div}^g$ is of Dirac type which suggest that equation (71) is symmetric hyperbolic. The disadvantage is that the solution might not be a 1-form. But this can be resoved by appropriate initial conditions.

Lemma 14 *Let (M, g) be a globally hyperbolic manifold and Σ a Cauchy hypersurface. If $\omega \in \Omega^*(M)$ is a solution of Eq. (71) with $\omega|_\Sigma \in T^*M$, then $\omega \in T^*M$ everywhere on M.*

Proof Because of $(d + \mathrm{div}^g)\omega = 0$ we get that ω is in the kernel of the Hodge Laplacian $\Delta^{HL} = (d + \mathrm{div}^g)^2$. Writing $\omega = \omega_1 + \ldots + \omega_{n+1}$ in components with k-forms ω_k, this implies that all ω_k satisfy the wave equation

$$\Delta^{HL}\omega_k = 0.$$

With $\omega|_\Sigma \in T^*M$ all ω_i except ω_1 vanish along Σ and hence, by Theorem 3, this wave equation implies that the ω_i's for $i > 1$ vanish everywhere. □

Next, as V is parallel, it annihilates the curvature tensor R^g, i.e., $V \lrcorner R^g = 0$. In particular,

$$V \lrcorner Ric^g = 0. \tag{73}$$

To evaluate this further, note that the vector field metric V defines a (non-orthogonal) splitting

$$TM = T\Sigma \oplus \mathbb{R}V \tag{74}$$

of bundles over M. We introduce the g- and V-dependent projection

$$\mathrm{pr}^{g,V}_{T\Sigma} : TM \to T\Sigma$$

onto the first factor in the splitting (74). Writing this explicitly as

$$\mathrm{pr}^{g,V}_{T\Sigma} = \mathrm{Id}_{TM} + \tfrac{1}{v}\left(g(T, .) - g(T, \mathrm{pr}_{T\Sigma}(.))\right) V,$$

where v is the function defined in Eq. (70), show that this projection is indeed g-dependent. Then Eq. (73) is equivalent to

$$Ric^g = Z \circ \mathrm{pr}^{g,V}_{T\Sigma}, \tag{75}$$

where Z is a symmetric bilinear form on $T\Sigma$, i.e. $Z \in \Gamma(\pi^*(T^*\Sigma \otimes T^*\Sigma))$, which is trivially extended to a symmetric bilinear form on $TM = T\Sigma \oplus \mathbb{R}V$.

Finally, in order to find the first order equation for Z, we observe that the second Bianchi identity implies that $(\nabla^g_V Ric^g) = 0$ and hence

$$(\nabla^g_V Z)(X, Y) = 0 \quad \text{for all } X, Y \in \pi^* T\Sigma. \tag{76}$$

Of course, for reasons described in Sect. 6.2, Eq. (75) cannot be symmetric hyperbolic as it contains the Ricci tensor. Hence we have to perform the hyperbolic reduction described in Sect. 6.4: for a positive function $\lambda \in C^\infty(\mathbb{R} \times \Sigma)$, we fix a background metric $\widetilde{g}$ on $\mathbb{R} \times \Sigma$ as in (59), which defines a 1-form $\eta = \eta^{g,\widetilde{g}}$ as in (53) and, instead of Eq. (71), consider the equation

$$\widetilde{Ric}^g = Ric^g + \mathrm{Sym}(\nabla^g \eta) = Z \circ \mathrm{pr}^{g,V}_{T\Sigma} + \mathrm{Sym}(\nabla^g \eta). \tag{77}$$

For Eqs. (71), (76) and (77) we obtain the analogon of Theorem 21.

Theorem 24 *Let (Σ, h) be a Riemannian manifold equipped with a vector field U such that $u := \sqrt{h(U, U)} > 0$ and satisfying the constraint (25) for a symmetric bilinear form* II*, a smooth function $f \in C^\infty(\Sigma)$, a smooth 1-form σ and a smooth bilinear form Q. Moreover, let λ be a smooth positive function on $\mathbb{R} \times \Sigma$. On $\mathbb{R} \times \Sigma$ fix the background metric $\widetilde{g} = -\lambda^2 dt^2 + h$ as in (59). Then on a neighbourhood M of $\{0\} \times \Sigma$ in $\mathbb{R} \times \Sigma$ there is a unique Lorentzian metric g with a unique lightlike vector field V and a unique bilinear form Z that satisfy equation (71) for $\omega = V^\flat$, Eqs. (76) and (77), and the initial conditions*

$$g|_\Sigma = \widetilde{g}|_\Sigma, \qquad g'|_{T\Sigma \times T\Sigma} = -2\lambda|_\Sigma \mathrm{II}, \qquad V|_\Sigma = \tfrac{u}{\lambda}\partial_t - U \tag{78}$$

as well as

$$g'(\partial_t, \partial_t)|_\Sigma = f, \qquad g'(\partial_t, .)|_\Sigma = \sigma, \qquad Z|_\Sigma = Q. \tag{79}$$

Moreover, (M, g) contains (Σ, h) as Cauchy hypersurface.

Proof As in the proof of Theorem 21, at each point in Σ we fix coordinates $(x^0 = t, x^1, \ldots, x^n)$ and show that in such coordinates Eqs. (71), (76) and (77) are equivalent to a symmetric hyperbolic system for a form ω, a metric g and a bilinear form Z.

We start with Eq. (75): as in the proof of Theorem 21 the initial conditions in (78) ensure that Eq. (76) is equivalent to the symmetric hyperbolic system

$$\begin{aligned} \partial_0 g_{\mu\nu} &= K_{\mu\nu}, \\ g^{00}\partial_0 K_{\mu\nu} + 2g^{0i}\partial_i K_{\mu\nu} + g^{ij}\partial_i L_{\mu\nu,j} &= H_{\mu\nu}, \\ -g_i{}^j \partial_0 L_{\mu\nu,j} + g_i{}^j \partial_j K_{\mu\nu} &= 0, \end{aligned} \tag{80}$$

which is the same as the system (56), with the difference that the function $F_{\mu\nu}$ in (56) has to be replaced by

$$H_{\mu\nu} = F_{\mu\nu} + \left(Z \circ \mathrm{pr}^{g,V}_{T\Sigma}\right)_{\mu\nu},$$

where V is defined by $g(V,.) = \omega_1$ for $\omega = \omega_1 + \ldots \omega_{n+1}$ with ω_k a k-form. Hence $H_{\mu\nu}$ depends only algebraically on the unknowns $g_{\alpha\beta}$, $\partial_\gamma g_{\alpha\beta}$, $Z_{\mu\nu}$ and $\omega = (\omega_1)_\alpha + \ldots + (\omega_{n+1})_{\alpha\ldots\beta}$.

Next, we consider Eq. (71) in coordinates and identify ω with a smooth map to $\mathbb{R}^{2^{n+1}}$ depending on the coordinates. The key here is that de Rham operator $d + \mathrm{div}^g = c \circ \nabla^g$ in (71), where c denotes the Clifford multiplication in (72), is of Dirac type. Hence, one can use the Clifford identity $c(X)\cdot c(Y) + c(Y)\cdot c(X) = -2g(X,Y)$ to show that equation (71) is of the form

$$A^\mu(x^\alpha, g_{\alpha\beta}, \omega)\partial_\mu \omega = b(x^\alpha, g_{\alpha\beta}, \partial_\gamma g_{\alpha\beta}, \omega), \tag{81}$$

where the matrices A^μ are symmetric. Along Σ, where $g = \widetilde{g}$, A^0 is a positive multiple of the identity, and hence A^0 is positive definite in a neighbourhood of Σ.

Finally, it is easy to see that Eq. (76), when written in local coordinates as

$$\partial_0 Z_{kl} - \tfrac{\lambda}{v} U^i \partial_i Z_{kl} = 2\Gamma^i_{0(k} Z_{l)i} - 2\tfrac{\lambda}{v} U^i \Gamma^j_{i(k} Z_{l)j},$$

is equivalent to an equation of the form

$$\partial_0 Z_{kl} + a^i(x^\alpha, g_{\alpha\beta}, \partial_\gamma g_{\alpha\beta}, (\omega_1)_\alpha)\partial_i Z_{kl} = b_{kl}(x^\alpha, g_{\alpha\beta}, \partial_\gamma g_{\alpha\beta}, (\omega_1)_\alpha, Z_{\alpha\beta}). \tag{82}$$

We conclude that the system comprising Eqs. (80)–(82) is a quasilinear first order symmetric hyperbolic system in the unknowns g, K, Z and ω, which has a unique solution for given initial conditions. Patching these together to a globally hyperbolic Lorentzian manifold (M, g) satisfying Eqs. (71), (75) and (76) for $V = \omega_1^\sharp$ works as before. Moreover, for ω we fix the initial condition

$$\omega|_\Sigma = \widetilde{g}(\tfrac{u}{\lambda}\partial_t - U, .) \in T^*M.$$

By Lemma 14, this initial condition implies that ω is in fact a 1-form on M and we have $V = \omega^\sharp$, which concludes the proof. □

After this result it remains to show that we can choose the remaining initial conditions (79) in a way that the associated solutions satisfy $\eta^{g,\widetilde{g}} = 0$ and $\nabla^g V = 0$. In particular to conclude $\nabla^g V = 0$ from $(d + \mathrm{div}^g)V^\sharp = 0$ seems a long shot, but as before, the right wave equation will help with this. This is the statement of the following proposition. Its proof in [62, Section 4] is substantially more involved than the proofs of Lemmas 9–11 above, and we do not attempt to present it here.

Proposition 8 *Let (Σ, h) be a Riemannian manifold and $\widetilde{g}$ be the background metric (59) on $\mathbb{R} \times \Sigma$. Moreover, let (M, g) be a Lorentzian manifold where M is an open neighbourhood of Σ, $\eta = \eta^{g,\widetilde{g}}$ be the 1-form defined in (53), V a lightlike vector field and Z a t-dependent bilinear form on Σ satisfying the equations*

$$Z = Ric^g + \mathrm{Sym}(\nabla^g \eta), \quad (d + \mathrm{div}^g)V^\flat = 0, \quad \nabla^g_V Z|_{T\Sigma \times T\Sigma} = 0, \quad V \lrcorner Z = 0.$$

(i) Then the tensor fields

$$\Phi := \left(\nabla^g V, \eta, \nabla^g_V \eta, (\nabla^g \eta)(V)\right), \qquad \Psi := \mathrm{div}^g (Z - \tfrac{\mathrm{tr}^g(Z)}{2} g),$$

satisfy a wave equation of the following form

$$\Delta_g \Phi = F(\Phi, \nabla^g \Phi, \Psi), \qquad \nabla^g_V \Psi = H(\Phi, \nabla^g \Phi), \tag{83}$$

where Δ_g is the Laplacian on tensor fields and F and H are suitable tensor fields. Moreover, when written in coordinates, the system (83) is equivalent to a linear first order symmetric hyperbolic system.

(ii) Assume that (Σ, h) and U satisfy the constraint conditions (24) and (25) and g, V, and Z satisfy the initial conditions along Σ,

$$g|_\Sigma = \widetilde{g}|_\Sigma, \quad g'|_\Sigma = -2\lambda \mathrm{II} + 2\lambda^2 \left(\lambda \mathrm{tr}^h(\mathrm{II}) - \lambda'\right) dt^2, \quad V|_\Sigma = \tfrac{u}{\lambda}\partial_t - U, \tag{84}$$

as well as

$$\begin{aligned} U \lrcorner Z|_\Sigma &= d\mathrm{tr}(W) - \mathrm{div}^h(W), \\ Z|_{U^\perp \times U^\perp} &= Ric^h - \mathrm{II}^2 + \mathrm{tr}(W)\mathrm{II} + \\ &\quad + \tfrac{1}{u^2}\left((U \lrcorner \mathrm{II})^2 - \mathrm{II}(U,U)\mathrm{II} - R^h(.,U,U,.)\right), \end{aligned} \tag{85}$$

Then $\Phi|_\Sigma = \Psi|_\Sigma = 0$.

The importance of the initial conditions (85) for Z becomes clear when comparing them to the constraint conditions for the Einstein equations in Lemma 6. This proposition is the key for the proof of the main result.

Theorem 25 *Let (Σ, h) be a Riemannian manifold equipped with a vector field U such that $u := \sqrt{h(U,U)} > 0$ and satisfying the constraint (25) for a symmetric bilinear form* II. *Moreover, let λ be a smooth positive function on $\mathbb{R} \times \Sigma$. On $\mathbb{R} \times \Sigma$ fix the background metric $\widetilde{g} = -\lambda^2 dt^2 + h$ as in (59). Then on a neighbourhood M of $\{0\} \times \Sigma$ in $\mathbb{R} \times \Sigma$ there is a unique Lorentzian metric g with a unique lightlike* *parallel vector field* V *such that* $\eta^{g,\widetilde{g}} = 0$ *and initial conditions*

$$g|_\Sigma = \widetilde{g}|_\Sigma, \qquad g'|_{T\Sigma \times T\Sigma} = -2\lambda|_\Sigma \mathrm{II}, \qquad V|_\Sigma = \tfrac{u}{\lambda}\partial_t - U$$

are satisfied. Moreover, (M, g) *is globally hyperbolic and contains* (Σ, h) *as Cauchy hypersurface.*

Proof We apply Theorem 24 in which we specify the initial conditions for f, σ and Q by Eqs. (84) and (85) as in (ii) of Proposition 8. Then (i) and (ii) in Proposition 8 imply $\Phi = \Psi = 0$ on all of M which in turn gives that $\eta^{g,\widetilde{g}} = 0$ as well as $\nabla^g V = 0$. In particular this implies that V is lightlike. As we have specified a full set of initial conditions for a symmetric hyperbolic system, g and V are unique. □

Remark 6 As explained in Remark 5, we can find a transformation fixing Σ such that the metric in Theorem 25 is of the form $g = -\widetilde{\lambda}^2 + h_t$ with a positive function $\widetilde{\lambda}$ such that $\widetilde{\lambda}|_\Sigma = \lambda|_\Sigma$.

7.3 Cauchy Problem for Parallel Lightlike Spinors

Now we will show that a Riemannian manifold (Σ, h) with imaginary W-Killing spinor of type I as defined in Definition 9 can always be extended to a Lorentzian manifold with parallel lightlike spinor. Since such a manifold satisfies the constraint conditions (24) and (25) for a parallel lightlike vector field (see Lemma 5), we can apply the results in the previous section and obtain a Lorentzian manifold (M, g) with parallel lightlike vector field. If (M, g) then admits a parallel spinor, it must be parallel translated along any direction that is transversal to Σ in M, e.g., parallel translated along the flow of V or of ∂_t.

Theorem 26 *Let* (Σ, h) *be a Riemannian manifold with imaginary* W*-Killing spinor* φ.

1. *If* (Σ, h), W *and* φ *are real analytic, then the real analytic globally hyperbolic Lorentzian manifold* (M, g) *obtained in Theorem 23 admits a unique real analytic parallel lightlike spinor field* ϕ *with* $\phi|_\Sigma = \varphi$ *and* $V = V_\phi$.
2. *If* (Σ, h), W *and* φ *are smooth, then the smooth globally hyperbolic Lorentzian manifold* (M, g) *obtained in Theorem 25 admits a unique smooth parallel lightlike spinor field* ϕ *with* $\phi|_\Sigma = \varphi$ *and* $V = V_\phi$.

Proof Let (M, g) be the globally hyperbolic Lorentzian manifold obtained from Theorem 23 or Theorem 25 and V be the parallel lightlike vector field on (M, g). Then we extend the spinor φ to a spinor on M by parallel transporting φ along the flow lines of V. This yields a real analytic (or smooth, depending on the assumptions) spinor field ϕ with $\nabla^S_V \phi = 0$. In order to show that also $\nabla^S_X \phi = 0$ for $X \in T\Sigma_t$, we consider the following section in $T^*\Sigma \otimes S$, where S is the spinor bundle of (M, g),

$$A(X) := \nabla^S_X \phi.$$

Then, using the fact that V is parallel, that $\nabla^S_V \phi = 0$, and the relation between the curvature R^S and R^g in Eq. (16), we compute

$$(\nabla^S_V A)(X) = R^S(V, X)\phi + \nabla^S_X \nabla^S_V \phi + \nabla^S_{[V,X]}\phi - \nabla^S_{\nabla^g_V X}\phi = R^g(V, X) \cdot \phi = 0.$$

This however is a linear first order PDE for A in Cauchy-Kowalevski form and in particular symmetric hyperbolic, so A is unique. The constraint equations and $\nabla^S_V \phi = 0$ however yield that $A|_\Sigma = 0$, which implies that $A = 0$ on all of M. The proof that V is the Dirac current of ϕ is straightforward and the details can be found in [16, Section 5]. □

Remark 7 This Theorem in the analytic setting was proven in [16] by the same method as presented here.

The result in the smooth setting was first proven by Lischewski [65], but without using the solution for the Cauchy problem for a parallel lightlike vector field presented in Sect. 7. Instead, Lischewski considers a Cauchy problem directly for the spinor and the metric by finding a PDE that corresponds to a first order symmetric hyperbolic system. Since the equation $\nabla^S \phi = 0$ does not have this property, the following system of equations is considered in [65],

$$Ric^g = f\,(V_\phi^\flat)^2, \qquad D^g \phi = 0, \qquad df(V_\phi) = 0, \tag{86}$$

where D^g is the Dirac operator of g. Then the the proof follows the one for the Einstein equations: first it is shown that these equations for the unknowns g and ϕ are first order quasilinear symmetric hyperbolic and hence have a unique solution, which then is shown to satisfy a wave equation that involves the square D^2 of the Dirac operator and the connection Laplacian Δ, and then the initial conditions are fixed so that the vanishing along the initial surface Σ is ensured.

8 Geometric Applications

In this section we will apply the results from the previous section to study Lorentzian holonomy reductions further. In Theorem 28 we also give an application our result has for Riemannian manifolds with imaginary W-Killing spinors, and in Remark 1 we formulate an open problem about flows of special Riemannian structures.

8.1 Applications to Lorentzian Holonomy

In this section we will use the results from Sect. 7 in order to construct globally hyperbolic Lorentzian manifolds with *prescribed screen holonomy* from Riemannian manifolds. On the one hand, from Theorem 17 we know how the Riemaniann manifolds satisfying the constraint conditions are constructed from families of Riemannian metrics, on the other hand we can use Theorem 25 to construct Lorentzian manifolds with special holonomy from them. We will now study the consequence for the family of Riemannian metrics that are implied by certain special holonomies.

The key concepts are the following:

Definition 12 Let (M, g) be a Lorentzian manifold of dimension $(n+2)$ with a subbundle $\mathbb{V}$ of parallel lightlike lines in the tangent bundle that is invariant under parallel transport. Then the bundle

$$\mathbb{S} := \mathbb{V}^{\perp}/\mathbb{V}$$

is called the *screen bundle* of (M, g). The connection on $\mathbb{S}$ defined by

$$\nabla^{\mathbb{S}}_X[Y] = \left[\nabla^g_X Y\right],$$

for $X \in TM$ and $Y \in \Gamma(\mathbb{V}^{\perp})$ is called *screen connection* and its holonomy group is called *screen holonomy group*. In the following we are only interested in the connected component of the screen holonomy group (or its Lie algebra), which we will simply call *screen holonomy*.

A subbundle $\mathbb{S} \subset TM$ is called *screen distribution* if $\mathbb{V}^{\perp} = \mathbb{V} \oplus \mathbb{S}$.

This is well defined as $\Gamma(\mathbb{V}^{\perp})$ and $\Gamma(\mathbb{V})$ are invariant under ∇^g. Since $\mathbb{V}$ is lightlike, we can also define a metric on $\mathbb{S}$,

$$g^{\mathbb{S}}([X], [Y]) := g(X, Y),$$

for $X, Y \in \mathbb{V}^{\perp}$. This metric is parallel with respect to $\nabla^{\mathbb{S}}$. As we mentioned in Sect. 3.2, the orthogonal component of the holonomy algebra of (M, g) is equal to the screen holonomy.

Exercise 7 Prove the last statement, i.e.,

$$\mathfrak{hol}(\mathbb{S}, \nabla^{\mathbb{S}}) \simeq \mathrm{proj}_{\mathfrak{so}(n)}(\mathfrak{hol}(M, g)).$$

According to Theorem 14 the screen holonomy is always the holonomy algebra of a Riemannian manifold. We will now establish a relation between the screen holonomy and the holonomy of the Riemannian metric in the families g_s in Theorem 17.

In general, the screen bundle is just a vector bundle over M, however in some situations it can be realised as a screen distribution. This is in particular the case when (M, g) admits a timelike unit vector field T. Then the screen bundle is realised as screen distribution

$$\mathbb{S} = T^{\perp} \cap \mathbb{V}^{\perp}.$$

Moreover, if (Σ, h) is a Riemannian manifold satisfying the constraint conditions (24) and (25) with a vector field U and (M, g) the Lorentzian manifold arising from it by Theorem 25, then by Remark 6 it is $T^{\perp}|_{\Sigma} = \partial_t^{\perp}|_{\Sigma} = T\Sigma$ and we can identify the screen distribution along Σ with

$$\mathbb{S}|_{\Sigma} = U^{\perp}.$$

For the screen connection along Σ we have

$$\nabla^{\mathbb{S}}_X Y|_{\Sigma} = \nabla^{\perp}_X Y, \qquad \text{for } Y \in \Gamma(\mathbb{S}|_{\Sigma}) = \Gamma(U^{\perp}),\ X \in T\Sigma, \tag{87}$$

where $\nabla^{\perp} = \mathrm{pr}_{U^{\perp}} \circ \nabla^h$ is the induced connection. Moreover if

$$(\Sigma, h) = (I \times N, \frac{1}{u^2} ds^2 + g_s)$$

is given as in Theorem 17, we can interpret $Y \in \Gamma(U^{\perp})$ as family $\{Y_s\}_{s \in I}$. Then we can compare the following vector bundles of the same rank

$$\begin{array}{ccccc} (TN, \nabla^{g_s}) & & (U^{\perp}, \nabla^{\perp}) & & (\mathbb{S}, \nabla^{\mathbb{S}}) \\ \downarrow & & \downarrow & & \downarrow \\ N & \subset & \Sigma & \subset & M. \end{array}$$

Next, recall that Lorentzian holonomy reductions from $\mathfrak{so}(1, n)$ to $\mathfrak{g} \ltimes \mathbb{R}^{n-1}$ with $\mathfrak{g} \subset \mathfrak{so}(n-1)$ are given by a parallel lightlike vector field V and parallel sections of

$$\otimes^{a,b}\mathbb{S} := \underbrace{\mathbb{S}^* \otimes \cdots \otimes \mathbb{S}^*}_{a \text{ times}} \otimes \underbrace{\mathbb{S} \otimes \cdots \otimes \mathbb{S}}_{b \text{ times}} \to M$$

such as a complex structure, a stable 3-form, etc. In this situation we have:

Proposition 9 *Let $(\Sigma, h) = (I \times N, \frac{1}{u^2} ds^2 + g_s)$ be a Riemannian manifold satisfying the constraints for a vector field U and (M, g) the Lorenzian manifold with parallel lightlike vector field V arising as solution of the Cauchy problem in Theorem 25. Then there is a 1–1 correspondence between*

(a) sections $\hat{\sigma}$ of the bundle $\mathbb{S}^{a,b} \to M$ with $\nabla^{\mathbb{S}} \sigma = 0$,

(b) *sections* σ *of the bundle* $\otimes^{a,b}U^{\perp} \to \Sigma$ *with* $\nabla^{\perp}\sigma = 0$, *and*
(c) *families* $(\sigma_s)_{s\in I}$ *of sections of* $\otimes^{a,b}TN \to N$ *with* $\nabla^{g_s}\sigma_s = 0$ *and*

$$\sigma_s' = \tfrac{1}{2}(g_s')^\sharp \bullet \sigma_s, \tag{88}$$

where the prime denotes the Lie derivative with respect to ∂_s, *the* $\sharp$ *is the dualisation with respect to* g_s *and* $\bullet$ *denotes the natural action of an endomorphism on tensors.*

Hence, $\mathfrak{hol}(\nabla^{\mathbb{S}}) = \mathrm{pr}_{\mathfrak{so}(n)}\mathfrak{hol}(M,g)$ *lies in the stabiliser of a tensor on* $\mathbb{S}$ *if and only if on* N *there is an induced s-dependent family of* g_s*-parallel tensors* σ_s *satisfying Eq. (88).*

Proof The equivalence of (b) and (c) is proved by just writing out the definition of $\nabla^{\perp}_{\partial_s}\sigma$ and $\nabla^{\perp}_X\sigma$ for $X \in TN$. The implication from (a) to (b) follows from the formula (87), so it remains to show the implication from (b) to (a). Then we extend $\sigma \in \Gamma(\otimes^{a,b}U^{\perp} \to \Sigma)$ to $\hat{\sigma} \in \Gamma(\otimes^{a,b}\mathbb{S} \to M)$ by parallel transport along the flow of V. Then the proof is similar to the one of Theorem 26: we define $A := \nabla^{\mathbb{S}}\sigma \in \Gamma(T^{\perp} \otimes \otimes^{a,b}\mathbb{S})$, show that $\nabla^{\mathbb{S}}_V\sigma = 0$, and observe that the equation $\nabla^{\mathbb{S}}_V A = 0$ is a linear symmetric hyperbolic system for A with initial condition $A|_\Sigma = 0$. Hence, not only $\nabla^{\mathbb{S}}_V\hat{\sigma} = 0$ but also $\nabla^{\mathbb{S}}_X\hat{\sigma} = 0$ for $X \in T\Sigma$. □

This result allows to describe the possible reductions of the (connected) screen holonomy geometrically in terms of the family g_s of Riemannian metrics on N. Later we will be interested in Lorentzian manifolds with parallel spinors, and from Theorem 16 we know that the screen holonomy has to admit a parallel spinor, i.e., has to be a product of SU(m), Sp(k), G_2, or Spin(7). As we have seen in Theorem 10, the corresponding geometric structures are special Kähler, hyper-Kähler, which are defined by (three) parallel complex structures J, as well as G_2 and Spin(7)-structures. A G_2-structure in dimension 7 is given by a parallel stable 3-form ϕ, i.e., a 3-form which has an open orbit in the 3-forms under the action of $\mathrm{GL}_7\mathbb{R}$, whereas a Spin(7)-structure in dimension 8 is given by a certain generic 4-form ψ. In both cases the forms define the metric, so in the next Theorem we will write we write $g_s = g_s(\phi_s)$ and $g_s = g_s(\psi_s)$ to indicate that for families of G_2 and Spin(7) structures, the metric g_s is defined algebraically in terms of a distinguished stable 3-form ϕ_s or a generic 4-form ψ_s, respectively. The explicit formulae can be found for example in [24, 52].

Theorem 27 *Let* (Σ, h, W, U) *be given as in* (26) *and* (27) *and let* (M, g) *be the Lorentzian manifold arising from this choice of initial data via Theorem 25 (for arbitrary choice for* λ*). Then* $G = \mathrm{pr}_{\mathrm{SO}(n)}\mathrm{Hol}(M,g) \subset \mathrm{SO}(n)$ *lies in the stabiliser of some tensor in* $T^{k,l}\mathbb{R}^n$ *if and only if there is an s-dependent and* ∇^{g_s}*-parallel family of tensor fields* σ_s *on* N, *of the same type and subject to the flow equation*

$$\sigma_s' = -\frac{1}{2}(g_s')^\sharp \bullet \sigma_s.$$

Here, the prime denotes the Lie derivative of a tensor with respect to ∂_s, e.g., $\sigma_s' := \mathscr{L}_{\partial_s}\sigma_s$, $(g')^\sharp\bullet$ denotes the natural action of the endomorphism $(g')^\sharp \in End(TN)$ on tensors in $T^{k,l}N$, and $\sharp$ indicates the dualisation with respect to g_s. Moreover:

1. *There are proper subgroups H_1 and H_2 of* $\mathrm{SO}(n)$ *such that $G \subset H_1 \times H_2$ if and only if there is a local metric splitting*

$$(N, g_s) \cong (N_1 \times N_2, g_s^1 + g_s^2) \tag{89}$$

   with $\mathrm{Hol}(N_i, g_s^i) \subset H_i$.
2. *if G is contained in one of* $\mathrm{SU}(m)$, $\mathrm{Sp}(k)$, G_2, $\mathrm{Spin}(7)$ *or trivial, this translates into the conditions for Riemannian special holonomy metrics from Table 1.*

For most of the cases *proof* of this Theorem follows easily from Proposition 9 with the exception of $\mathrm{SU}(m)$. Here a lengthy computation in [62, Section 7] shows how the condition $\mathrm{Hol}(M, g) \subset \mathrm{SU}(m)$ translates into $\mathrm{div}^{g_s}(g_s') = 0$.

Theorems 25 and 27 provide a construction principle for Lorentzian manifolds with reduced screen holonomy as seen in the following simple example.

Example 11 Consider the following warped product manifold ($\Sigma = I \times N, h := h = ds^2 + f(s)g_0$) with (N, g_0) a Ricci-flat simply connected Riemannian manifold with irreducible special holonomy, i.e., $\mathrm{Hol}(N, h_0) \in \{\mathrm{SU}(m), \mathrm{Sp}(k), \mathrm{G}_2, \mathrm{Spin}(7)\}$ or trivial. Then(Σ, h) satisfies the constraint and Theorem 25 can be applied to it, obtaining a Lorentzian manifold with the corresponding screen holonomy $\mathrm{Hol}(N, h_0)$ by Theorem 27.

Problem 1 (Open Problems About Flows of Special Riemannain Structures) We do not know whether the flow equation (88) on the parallel tensors and the metric in Table 1 defining the holonomy reductions is an *extra* condition or if it

Table 1 Equivalent characterisation of special screen holonomy for (M, g) in terms of flow equations for tensors on N

$\dim(N)$	Geometric structures on N	$\mathrm{Hol}(M, g) \subset$
$2m$	$(N, \omega_s, J_s, g_s = \omega_s(J_s\cdot, \cdot))$ Ricci-flat Kähler, $J_s' = -\frac{1}{2}(g_s')^\sharp \bullet J_s,\ \mathrm{div}^{g_s}(g_s') = 0$	$\mathrm{SU}(m) \ltimes \mathbb{R}^{2m}$
$4k$	$(N, \omega_s^i, J_s^i, g_s = \omega_s^i(J_s^i\cdot, \cdot))_{i=1,2,3}$ hyper-Kaehler, $(J^i)_s' = -\frac{1}{2}(g_s')^\sharp \bullet J_s^i$	$\mathrm{Sp}(k) \ltimes \mathbb{R}^{4k}$
7	$(N, \phi_s \in \Omega^3(N)), g_s = g_s(\phi_s)$ G_2 metrics, $\phi_s' = -\frac{1}{2}(g_s')^\sharp \bullet \phi_s$	$\mathrm{G}_2 \ltimes \mathbb{R}^7$
8	$(N, \psi_s \in \Omega^4(N), g_s = g_s(\psi_s))$ Spin(7) metrics, $\psi_s' = -\frac{1}{2}(g_s')^\sharp \bullet \psi_s$	$\mathrm{Spin}(7) \ltimes \mathbb{R}^8$
n	g_s flat	$\mathbb{R}^n$

is automatically satisfied. This question is suggested by the following observation: when proving the first point (1) in Theorem 27 one shows that a parallel volume form, i.e., the parallel object that defines the splitting, always satisfies the flow equation (88).

The question can be formulated, for example, in the case of G_2-structures as follows: given a family of holonomy G_2-metrics h_s, does there exist a family of stable 3-forms ϕ_s, such that $h_s = h_s(\phi_s)$ and such that ϕ_s satisfies

$$\phi_s' + \frac{1}{2}(g_s')^\sharp \bullet \phi_s = 0\ ? \tag{90}$$

Since the tangent space at a stable three form ϕ splits under G_2 into three irreducible components

$$\begin{aligned} \mathbb{R} \oplus Sym_0^2(\mathbb{R}^7) \oplus \mathbb{R}^7 &\simeq \Lambda^3 \\ (r, S, X) &\mapsto r\phi + S^\sharp \bullet \phi + X \lrcorner (*\phi), \end{aligned}$$

where $*$ is the Hodge star operator, it follows that

$$\phi' = S^\sharp \bullet \phi + X \lrcorner (*\phi),$$

for a family of symmetric bilinear forms, whereas the associated metric satisfies $g' = 2S$, see [28, 53, 54]. Hence, for the curve ϕ_t the equation (90) is equivalent to the condition

$$\phi' \in \mathbb{R} \oplus Sym_0(\mathbb{R}^7),$$

i.e., that ϕ' has no $\mathbb{R}^7$-component in the decomposition. We do not know if this can always be achieved for any curve h_t of holonomy G_2-metrics, and we leave this as an *open problem*.

In the case of Kähler structures the situation is similar. It can be shown [62, Section 7] that the flow equation (88) for a Kähler form ω is equivalent to

$$\omega_s' \in \Lambda^{1,1}(N, J_s), \tag{91}$$

where $\Lambda^{1,1}(N, J_s)$ denotes the 2-forms that are holomorphic with respect to the parallel complex structure J_s. Again we leave it as an *open question* whether or not for a given family of Kähler metrics there are compatible parallel complex structures J_s and Kähler forms ω_s such that the relation (91) is satisfied.

The answers to these questions will have consequences for the results we will give now.

8.2 Applications to Spinor Field Equations

In this last section we want to apply the previous results, in particular the one in Theorem 26 to spinor field equations: to the equation for a parallel spinor on a Lorentzian manifold and that for an imaginary W-Killing spinor on a Riemannian manifold.

We start with the problem of a (local) classification of Riemannian manifolds with imaginary W-Killing spinor. This means that we change our perspective slightly from focussing on the Lorentzian manifold—which will now appear as a tool—to focussing on the Riemannian manifold, which so far appeared only as initial condition. Clearly, condition (31) for an imaginary W-Killing spinor arises as a generalisation of the equation for imaginary Killing spinors, for which $W = \frac{\mathrm{i}}{2}\,\mathrm{Id}$, see [14]. Moreover, solutions to equation (31) are the counterpart to real generalised Killing spinors which have been in the focus of recent research, for example in [3, 5]. A Riemannian manifold (Σ, h) with imaginary W-Killing spinor of type I (see Definition 10) satisfies the constraint conditions (24) and (25) for a vector field (Lemma 5). Hence, applying Theorem 17 gives a generalisation of results from [9, 10], see also [14], where it is shown that in the complete case and for $W = f\,\mathrm{Id}$, (Σ, h) is necessarily isometric to a warped product (see Example 8). Using the Lorentzian manifold (M, g) obtained by the Cauchy problem as a tool, with the results from the previous sections we can even say more.

Theorem 28 *Let (Σ, h) be a Riemannian spin manifold admitting an imaginary W-Killing spinor φ of type I. Then:*

1. *(Σ, h) is locally isometric to*

$$(\Sigma, h) \simeq \left(I \times N_1 \times \ldots \times N_k, h = \frac{1}{u^2} ds^2 + g_s^1 + \ldots + g_s^k \right) \tag{92}$$

   for Riemannian manifolds (N_i, g_s^i) of dimension n_i, $u = ||\varphi||^2$, I an interval, and under this isometry W is given by (27). *Moreover, for each $i = 1, \ldots, k$, each h_s^i is a family of special holonomy metrics to which exactly one of the cases of Table 1 applies.*
2. *If (Σ, h) is simply connected and the vector field $\frac{1}{u_\varphi^2} U_\varphi$ is complete, the isometry in* (92) *is global with $I = \mathbb{R}$.*
3. *Conversely, every Riemannian manifold (Σ, h) of the form* (92) *with $I \in \{S^1, \mathbb{R}\}$, where u is any positive function and (N_i, g_s^i) are families of special holonomy metrics subject to the flow equations in Table 1, is spin and admits an imaginary W-Killing spinor φ of type I.*

The *proof* follows from several results above: if (Σ, h) admits an imaginary W-Killing spinor of type I, byTheorem 26 there is a Lorentzian manifold (M, g) which admits a lightlike parallel spinor field (and hence a lightlike parallel vector field). By Theorem 16, the connected screen holonomy of (M, g) is a product of the groups listed in Table 1. Then the results in Theorem 27 then imply the existence of the local isometry to a metric (92). The statements in (2) and (3) also follow from results in Theorem 17 and using the Lorentzian manifold arising from the Cauchy problem.

Finally, we give a local normal form for Lorentzian metrics admitting a parallel *lightlike* spinor fields.

Theorem 29 *Let (M, g) be a Lorentzian manifold admitting a parallel lightlike spinor field. Then (M, g) is locally isometric to*

$$(M, g) \cong (\mathbb{R} \times \mathbb{R} \times N_1 \times \ldots \times N_m,\ 2dvdw + g_w^1 + \ldots + g_w^m), \tag{93}$$

for some integer m, manifolds N_i for $i = 1, \ldots, m$ where each g_w^i is a w-dependent family of Riemannian metrics on N_i to which exactly one of the cases in Table 1 applies. Conversely, every manifold as in (93) satisfying these conditions admits a parallel lightlike spinor.

Again, this theorem follows from the previous results and by introducing new coordinates $v = -t + s$ and $w = t + s$ in which the metric in (93) is of the form

$$g = -dt^2 + ds^2 + g_{t+s} =: -dt^2 + h_t. \tag{94}$$

Note that the normal forms in Theorem 29 need not be the most general ones. For example, for signature $(1, 10)$, in [27] it is shown that a term $H_w dw^2$, where H_w is an arbitrary function not depending on v can be added to (93). However, the analysis of normal forms for metrics with parallel spinor in [27] rests on the known orbit structure of the action of $\mathrm{Spin}(1, n)$ in low dimensions whereas Theorem 29 covers all dimensions.

References

1. D.V. Alekseevsky, B.N. Kimmelfeld, Structure of homogeneous Riemannian spaces with zero Ricci curvature. Funct. Anal. Appl. **9**(2), 5–11 (1975)
2. D. Alekseevsky, V. Cortés, A. Galaev, T. Leistner, Cones over pseudo-Riemannian manifolds and their holonomy. J. Reine Angew. Math. **635**, 23–69 (2009)
3. B. Ammann, A. Moroianu, S. Moroianu, The Cauchy problems for Einstein metrics and parallel spinors. Commun. Math. Phys. **320**(1), 173–198 (2013)
4. C. Bär, Real Killing spinors and holonomy. Commun. Math. Phys. **154**(3), 509–521 (1993)
5. C. Bär, P. Gauduchon, A. Moroianu, Generalized cylinders in semi-Riemannian and Spin geometry. Math. Z. **249**(3), 545–580 (2005)
6. C. Bär, N. Ginoux, F. Pfäffle, *Wave Equations on Lorentzian Manifolds and Quantization*. ESI Lectures in Mathematics and Physics (European Mathematical Society, Zürich, 2007)
7. R. Bartnik, J. Isenberg, The constraint equations, in *The Einstein Equations and the Large Scale Behavior of Gravitational Fields* (Birkhäuser, Basel, 2004), pp. 1–38
8. H. Baum, *Spin-Strukturen und Dirac-Operatoren über pseudoriemannschen Mannigfaltigkeiten*. Teubner-Texte zur Mathematik, vol. 41 (Teubner-Verlagsgesellschaft, Leipzig, 1981)

9. H. Baum, Complete Riemannian manifolds with imaginary Killing spinors. Ann. Global Anal. Geom. **7**(3), 205–226 (1989)
10. H. Baum, Odd-dimensional Riemannian manifolds with imaginary Killing spinors. Ann. Global Anal. Geom. **7**(2), 141–153 (1989)
11. H. Baum, A remark on the spectrum of the Dirac operator on pseudo-Riemannian spin maniofolds. Technical Report 136, SFB 288-Preprint, 1994
12. H. Baum, *Eichfeldtheorie. Eine Einführung in die Differentialgeometrie auf Faserbündeln*, 2nd revised edition (Springer Spektrum, Heidelberg, 2014)
13. H. Baum, O. Müller, Codazzi spinors and globally hyperbolic manifolds with special holonomy. Math. Z. **258**(1), 185–211 (2008)
14. H. Baum, T. Friedrich, R. Grunewald, I. Kath, Twistors and Killing spinors on Riemannian manifolds. Teubner-Texte zur Mathematik [Teubner Texts in Mathematics], vol. 124 (B. G. Teubner Verlagsgesellschaft mbH, Stuttgart, 1991)
15. H. Baum, K. Lärz, T. Leistner, On the full holonomy group of Lorentzian manifolds. Math. Z. **277**(3–4), 797–828 (2014)
16. H. Baum, T. Leistner, A. Lischewski, Cauchy problems for Lorentzian manifolds with special holonomy. Differ. Geom. Appl. **45**, 43–66 (2016)
17. Ya. V. Bazaikin, Globally hyperbolic Lorentzian spaces with special holonomy groups. Siberian Math. J. **50**(4), 567579 (2009)
18. L. Bérard-Bergery, A. Ikemakhen, On the holonomy of Lorentzian manifolds, in *Differential Geometry: Geometry in Mathematical Physics and Related Topics (Los Angeles, CA, 1990). Proc. Sympos. Pure Math.*, vol. 54 (American Mathematical Society, Providence, RI, 1993), pp. 27–40
19. M. Berger, Sur les groupes d'holonomie homogène des variétés à connexion affine et des variétés riemanniennes. Bull. Soc. Math. France **83**, 279–330 (1955)
20. M. Berger, Les espaces symétriques noncompacts. Ann. Sci. École Norm. Sup. (3) **74**, 85–177 (1957)
21. A.N. Bernal, M. Sánchez, Smoothness of time functions and the metric splitting of globally hyperbolic spacetimes. Commun. Math. Phys. **257**(1), 43–50 (2005)
22. A.L. Besse, *Einstein Manifolds* (Springer, Berlin/Heidelberg/New York, 1987)
23. N. Bezvitnaya, Lightlike foliations on Lorentzian manifolds with weakly irreducible holonomy algebra (2005). Preprint. arXiv:math.DG/0506101
24. R.L. Bryant, Metrics with exceptional holonomy. Ann. Math. (2) **126**(3), 525–576 (1987)
25. R.L. Bryant, Classical, exceptional, and exotic holonomies: a status report. Séminaires Conrès Soc. Math. France **1**, 93–165 (1996)
26. R.L. Bryant, Recent advances in the theory of holonomy. Séminaire Bourbaki, Asterisque **51**, 525–576 (1999)
27. R.L. Bryant, Pseudo-Riemannian metrics with parallel spinor fields and vanishing Ricci tensor, in *Global Analysis and Harmonic Analysis (Marseille-Luminy, 1999)*. Sémin. Congr., vol. 4 (Société Mathématique de France, Paris, 2000), pp. 53–94
28. R.L. Bryant, Some remarks on G_2-structures, in *Proceedings of Gökova Geometry-Topology Conference 2005* (Gökova Geometry/Topology Conference (GGT), Gökova, 2006), pp. 75–109
29. M. Cahen, N. Wallach, Lorentzian symmetric spaces. Bull. Am. Math. Soc. **79**, 585–591 (1970)
30. A.M. Candela, J.L. Flores, M. Sanchez, On general plane fronted waves. Gen. Relat. Gravit. **35**(4), 631–649 (2003)
31. E. Cartan, La Geometrie des Espaces de Riemann. Memorial des Sciences Mathematiques, Fasc. IX, Ch. VII, Sec. II (1925)
32. E. Cartan, Sur une classe remarquable d'espaces Riemann. Bull. Soc. Math. France **54**, 214–264 (1926)
33. E. Cartan, Les groups d'holonomie des espaces généralisés. Acta Math. **48**, 1–42 (1926)
34. G. Darmois, *Les equations de la gravitation Einsteinienne*. Memorial des Sciences Mathématiques, vol. XXV (Gauthier Villars, Paris, 1927)

35. G. de Rham, Sur la reductibilité d'un espace de Riemann. Comment. Math. Helv. **26**, 328–344 (1952)
36. A.J. Di Scala, C. Olmos, The geometry of homogeneous submanifolds of hyperbolic space. Math. Z. **237**(1), 199–209 (2001)
37. J.L. Flores, M. Sanchez, Causality and conjugated points in general plane waves. Class. Quant. Gravity **20**, 2275–2291 (2003)
38. G.B. Folland, *Introduction to Partial Differential Equations*, 2nd edn. (Princeton University Press, Princeton, NJ, 1995)
39. Y. Fourès-Bruhat, Théorème d'existence pour certains systèmes d'équations aux dérivées partielles non linéaires. Acta Math. **88**, 141–225 (1952)
40. F.G. Friedlander, *The Wave Equation on Curved Space Time* (Cambridge University Press, Cambridge, MA, 1975)
41. T. Friedrich, *Dirac Operators in Riemannian Geometry. Transl. from the German by Andreas Nestke* (American Mathematical Society, Providence, RI, 2000)
42. H. Friedrich, A. Rendall, The Cauchy problem for the Einstein equations, in *Einstein's Field Equations and Their Physical Implications*. Lecture Notes in Physics, vol. 540 (Springer, Berlin, 2000), pp. 127–223
43. A.S. Galaev, Isometry groups of Lobachevskian spaces, similarity transformation groups of Euclidean spaces and Lorentzian holonomy groups. Rend. Circ. Mat. Palermo (2) Suppl. **79**, 87–97 (2006)
44. A.S. Galaev, Metrics that realize all Lorentzian holonomy algebras. Int. J. Geom. Methods Mod. Phys. **3**(5–6), 1025–1045 (2006)
45. A.S. Galaev, Holonomy groups of Lorentzian manifolds. Russ. Math. Surv. **70**(2), 249–298 (2015)
46. A.S. Galaev, T. Leistner, Holonomy groups of Lorentzian manifolds: classification, examples, and applications, in *Recent Developments in Pseudo-Riemannian Geometry*. ESI Lectures in Mathematics and Physics, pp. 53–96 (European Mathematical Society, Zürich, 2008)
47. S. Gallot, Équations différentielles caractéristiques de la sphère. Ann. Sci. École Norm. Sup. (4) **12**(2), 235–267 (1979)
48. R. Geroch, Domain of dependence. J. Math. Phys. **11**, 437–449 (1970)
49. G.S. Hall, D.P. Lonie, Holonomy groups and spacetimes. Classical Quantum Gravity **17**(6), 1369–1382 (2000)
50. S. Helgason, *Differential Geometry, Lie Groups, and Symmetric Spaces*. Graduate Studies in Mathematics, vol. 34 (American Mathematical Society, , Providence, RI, 2001)
51. D.D. Joyce, *Compact Manifolds with Special Holonomy*. Oxford Mathematical Monographs (Oxford Univerity Press, Oxford, 2000)
52. D.D. Joyce, *Riemannian Holonomy Groups and Calibrated Geometry*. Oxford Graduate Texts in Mathematics, vol. 12 (Oxford University Press, Oxford, 2007)
53. S. Karigiannis, Deformations of G_2 and $Spin(7)$ structures. Can. J. Math. **57**(5), 1012–1055 (2005)
54. S. Karigiannis, Flows of G_2-structures. I. Q. J. Math. **60**(4), 487–522 (2009)
55. M. Karoubi, Algèbres de Clifford et K-théorie. Ann. Sci. École Norm. Sup. (4) **1**, 161–270 (1968)
56. N. Koiso, Hypersurfaces of Einstein manifolds. Ann. Sci. École Norm. Sup. (4) **14**(4), 433–443 (1981)
57. K. Lärz, Global Aspects of Holonomy in Pseudo-Riemannian Geometry. PhD thesis, Humboldt-Universität zu Berlin, 2011. Available at http://edoc.hu-berlin.de/dissertationen/
58. H.B. Lawson, M.-L. Michelsohn, *Spin Geometry* (Princeton University Press, Princeton, 1989)
59. T. Leistner, Berger algebras, weak-Berger algebras and Lorentzian holonomy, 2002. SFB 288-Preprint no. 567, ftp://ftp-sfb288.math.tu-berlin.de/pub/Preprints/preprint567.ps.gz

60. T. Leistner, Holonomy and parallel spinors in Lorentzian geometry. PhD thesis, Berlin: Logos-Verlag; Berlin: Humboldt-Univ., Mathematisch-Naturwissenschaftliche Fakultät II (Dissertation). xii, 178 p. , 2003
61. T. Leistner, On the classification of Lorentzian holonomy groups. J. Differ. Geom. **76**(3), 423–484 (2007)
62. T. Leistner, A. Lischewski, Hyperbolic evolution equations, Lorentzian holonomy, and Riemannian generalised Killing spinors. J. Geom. Anal. (2017). https://doi.org/10.1007/s12220-017-9941-x
63. T. Leistner, D. Schliebner, Completeness of compact Lorentzian manifolds with abelian holonomy. Math. Ann. **364**(3–4), 1469–1503 (2016)
64. A. Lichnerowicz, Problèmes globaux en mécanique relativiste. Actual. Sci. Ind. 833. Exposés de géométrie. XII. Hermann et Cie, Paris, 1939
65. A. Lischewski, The Cauchy problem for parallel spinors as first-order symmetric hyperbolic system (March 2015). Preprint. arXiv:1503.04946
66. E. Minguzzi, M. Sánchez, The causal hierarchy of spacetimes, in *Recent Developments in Pseudo-Riemannian Geometry*. ESI Lectures in Mathematics and Physics (European Mathematical Society, Zürich, 2008), pp. 299–358
67. A. Moroianu, U. Semmelmann, Parallel spinors and holonomy groups. J. Math. Phys. **41**(4), 2395–2402 (2000)
68. B. O'Neill, *Semi-Riemannian Geometry* (Academic, Cambridge, 1983)
69. H.-B. Rademacher, Generalized Killing spinors with imaginary Killing function and conformal Killing fields, in *Global Differential Geometry and Global Analysis (Berlin, 1990)*. Lecture Notes in Mathematics, vol. 1481 (Springer, Berlin, 1991), pp. 192–198
70. H. Ringström, *The Cauchy Problem in General Relativity*. ESI Lectures in Mathematics and Physics (European Mathematical Society, Zürich, 2009)
71. S.M. Salamon, *Riemannian Geometry and Holonomy Groups*. Pitmann Research Lecture Notes, vol. 201 (Longman Scientific & Technical, Harlow, 1989)
72. J.F. Schell, Classification of 4-dimensional Riemannian spaces. J. Math. Phys. **2**, 202–206 (1960)
73. R. Shaw, The subgroup structure of the homogeneous Lorentz group. Quart. J. Math. Oxford **21**, 101–124 (1970)
74. M.E. Taylor, *Partial Differential Equations III. Nonlinear Equations*. Applied Mathematical Sciences, vol. 117, 2nd edn. (Springer, New York, 2011)
75. M.Y. Wang, Parallel spinors and parallel forms. Ann. Global Anal. Geom. **7**(1), 59–68 (1989)
76. M.Y. Wang, On non-simply connected manifolds with non-trivial parallel spinors. Ann Global Anal. Geom. **13**, 31–42 (1995)
77. H. Wu, On the de Rham decomposition theorem. Illinois J. Math. **8**, 291–311 (1964)

Geometric Flow Equations

Oliver C. Schnürer

Contents

Abstract In this minicourse, we study hypersurfaces that solve geometric evolution equations. More precisely, we investigate hypersurfaces that evolve with a normal velocity depending on a curvature function like the mean curvature or Gauß

O. C. Schnürer (✉)
Fachbereich Mathematik und Statistik, Universität Konstanz, Konstanz, Germany
e-mail: Oliver.Schnuerer@uni-konstanz.de

V. Cortés et al. (eds.), *Geometric Flows and the Geometry of Space-time*, Tutorials, Schools, and Workshops in the Mathematical Sciences,
https://doi.org/10.1007/978-3-030-01126-0_2

curvature. In three lectures, we address

- hypersurfaces, principal curvatures and evolution equations for geometric quantities like the metric and the second fundamental form.
- the convergence of convex hypersurfaces to round points. Here, we will also show some computer algebra calculations.
- the evolution of graphical hypersurfaces under mean curvature flow.

1 Overview and Plan for the Summer School

We consider flow equations that deform hypersurfaces according to their curvature.

If $X_0 : M^n \to \mathbb{R}^{n+1}$ is an embedding of an n-dimensional manifold, we can define principal curvatures $(\lambda_i)_{1\le i\le n}$ and a normal vector ν. We deform the embedding vector X according to

$$\begin{cases} \frac{d}{dt}X = -F\nu, \\ X(\cdot, 0) = X_0, \end{cases}$$

where F is a symmetric function of the principal curvatures, e.g. the mean curvature $H = \lambda_1 + \cdots + \lambda_n$. In this way, we obtain a family $X(\cdot, t)$ of embeddings and study their behaviour especially near singularities and for large times. We consider hypersurfaces that contract to a point in finite time and, after appropriate rescaling, to a round sphere. Graphical solutions are shown to exist for all times or to disappear to infinity.

Classical results in this direction were obtained by Huisken [20] and Ecker and Huisken [11] for mean curvature flow.

Remark 1

(i) We will use geometric flow equations as a tool to deform a manifold.
(ii) The flow equations considered are parabolic equations like the heat equation.
(iii) In order to control the behaviour of the flow, we will look for properties of the manifold that are preserved under the flow. For that purpose, we will also look for quantities that are monotone and have geometric significance, i.e. their boundedness implies geometric properties of the evolving manifold.

We wish to thank Ben Lambert and Wolfgang Maurer for corrections and Wolfgang Maurer for carefully preparing the figures.

1.1 Plan for the Summer School

These notes first cover some necessary background material. We will then derive evolution equations for geometric quantities and study two geometric problems. More precisely, our plan is to study the following:

- Geometric prerequisites and evolution equations of geometric quantities.
- Convex surfaces contracting to a round point and an estimate for Gauß curvature flow, Theorem 6, measuring the deviation from being umbilic.
- Mean curvature flow of complete graphs and local C^1-bounds, Theorem 16.

2 Differential Geometry of Submanifolds

We will only consider hypersurfaces in Euclidean space.

We use $X = X(x, t) = (X^\alpha)_{1\le\alpha\le n+1}$ to denote the time-dependent embedding vector of a manifold M^n into $\mathbb{R}^{n+1}$ and $\frac{d}{dt}X = \dot{X}$ for its total time derivative. Set $M_t := X(M, t) \subset \mathbb{R}^{n+1}$. We will often identify an embedded manifold with its image. We will assume that X is smooth. Assume furthermore that M^n is smooth, orientable, connected, complete and $\partial M^n = \emptyset$. We choose $\nu = \nu(x) = (\nu^\alpha)_{1\le\alpha\le n+1}$ to be the outer (or downward pointing) unit normal vector to M_t at $x \in M_t$. The embedding $X(\cdot, t)$ induces at each point on M_t a metric $(g_{ij})_{1\le i,\, j\le n}$ and a second fundamental form $(h_{ij})_{1\le i,\, j\le n}$. Let (g^{ij}) denote the inverse of (g_{ij}). These tensors are symmetric. The principal curvatures $(\lambda_i)_{1\le i\le n}$ are the eigenvalues of the second fundamental form with respect to that metric. That is, at $p \in M$, for each principal curvature λ_i, there exists $0 \neq \xi \in T_pM \cong \mathbb{R}^n$ such that

$$\lambda_i \sum_{l=1}^n g_{kl}\xi^l = \sum_{l=1}^n h_{kl}\xi^l \quad \text{or, equivalently,} \quad \lambda_i\xi^l = \sum_{k,r=1}^n g^{lk}h_{kr}\xi^r.$$

As usual, eigenvalues are listed according to their multiplicity. A hypersurface is called strictly convex, if all principal curvatures are strictly positive. The inverse of the second fundamental form is denoted by $\left(\tilde{h}^{ij}\right)_{1\le i,\, j\le n}$.

Latin indices range from 1 to n and refer to geometric quantities on the hypersurface, Greek indices range from 1 to $n+1$ and refer to components in the ambient space $\mathbb{R}^{n+1}$. In $\mathbb{R}^{n+1}$, we will always choose Euclidean coordinates. We use the Einstein summation convention for repeated upper and lower indices. Latin indices are raised and lowered with respect to the induced metric or its inverse $\left(g^{ij}\right)$, for Greek indices we use the flat metric $(\overline{g}_{\alpha\beta})_{1\le\alpha,\beta\le n+1} = (\delta_{\alpha\beta})_{1\le\alpha,\beta\le n+1}$ of $\mathbb{R}^{n+1}$. So the defining equation for the principal curvatures becomes $\lambda_i g_{kl}\xi^l = h_{kl}\xi^l$.

Denoting by $\langle\cdot, \cdot\rangle$ the Euclidean scalar product in $\mathbb{R}^{n+1}$, we have

$$g_{ij} = \left\langle X_{,i}, X_{,j}\right\rangle = X^\alpha_{,i}\delta_{\alpha\beta}X^\beta_{,j},$$

where we used indices, preceded by commas, to denote partial derivatives. We write indices, preceded by semi-colons, e.g. $h_{ij;k}$ or $v_{;k}$, to indicate covariant differentiation with respect to the induced metric. Later, we will also drop the commas and semi-colons, if the meaning is clear from the context. We set $X^\alpha_{;i} \equiv X^\alpha_{,i}$

and

$$X^{\alpha}_{;\,ij} = X^{\alpha}_{,\,ij} - \Gamma^k_{ij} X^{\alpha}_{,\,k}, \tag{1}$$

where

$$\Gamma^k_{ij} = \tfrac{1}{2} g^{kl} (g_{il,\,j} + g_{jl,\,i} - g_{ij,\,l})$$

are the Christoffel symbols of the metric (g_{ij}). Therefore, $X^{\alpha}_{;ij}$ becomes a tensor.

The Gauß formula relates covariant derivatives of the position vector to the second fundamental form and the normal vector

$$X^{\alpha}_{;\,ij} = -h_{ij} \nu^{\alpha}. \tag{2}$$

The Weingarten equation allows to compute derivatives of the normal vector

$$\nu^{\alpha}_{;\,i} = h^k_i X^{\alpha}_{;\,k}. \tag{3}$$

We can use the Gauß formula (2) or the Weingarten equation (3) to compute the second fundamental form.

Symmetric functions of the principal curvatures are well-defined, we will use the mean curvature $H = \lambda_1 + \ldots + \lambda_n$, the square of the norm of the second fundamental form $|A|^2 = \lambda_1^2 + \ldots + \lambda_n^2$, $\operatorname{tr} A^k = \lambda_1^k + \ldots + \lambda_n^k$, and the Gauß curvature $K = \lambda_1 \cdot \ldots \cdot \lambda_n$. It is often convenient to choose coordinate systems such that, at a fixed point, the metric tensor equals the Kronecker delta, $g_{ij} = \delta_{ij}$, and (h_{ij}) is diagonal, $(h_{ij}) = \operatorname{diag}(\lambda_1, \ldots, \lambda_n)$, e.g.

$$\sum \lambda_k h^2_{ij;k} = \sum_{i,\,j,\,k=1}^{n} \lambda_k h^2_{ij;k} = h^{kl} h^i_{j;\,k} h^j_{i;\,l} = h_{rs} h_{ij;\,k} h_{ab;\,l} g^{ia} g^{jb} g^{rk} g^{sl}.$$

Whenever we use this notation, we will also assume that we have fixed such a coordinate system.

A normal velocity F can be considered as a function of $(\lambda_1, \ldots, \lambda_n)$ or $(h_{ij},\, g_{ij})$. If $F(\lambda_i)$ is symmetric and smooth, then $F(h_{ij}, g_{ij})$ is also smooth [17, Theorem 2.1.20]. We set $F^{ij} = \frac{\partial F}{\partial h_{ij}}$, $F^{ij,\,kl} = \frac{\partial^2 F}{\partial h_{ij} \partial h_{kl}}$. Note that in coordinate systems with diagonal h_{ij} and $g_{ij} = \delta_{ij}$ as mentioned above, F^{ij} is diagonal. For $F = |A|^2$, we have $F^{ij} = 2h^{ij} = 2\lambda_i g^{ij}$, and for $F = K^{\alpha}$, $\alpha > 0$, we have $F^{ij} = \alpha K^{\alpha} \tilde{h}^{ij} = \alpha K^{\alpha} \lambda_i^{-1} g^{ij}$.

The Gauß equation expresses the Riemannian curvature tensor of the hypersurface in terms of the second fundamental form

$$R_{ijkl} = h_{ik} h_{jl} - h_{il} h_{jk}. \tag{4}$$

As we use only Euclidean coordinate systems in $\mathbb{R}^{n+1}$, $h_{ij;\,k}$ is symmetric in all three indices according to the Codazzi equations.

The Ricci identity allows to interchange covariant derivatives. We will use it for the second fundamental form

$$h_{ik;\,lj} = h_{ik;\,jl} + h_k^a R_{ailj} + h_i^a R_{aklj}. \tag{5}$$

For tensors A and B, $A_{ij} \geq B_{ij}$ means that $(A_{ij} - B_{ij})$ is positive semi-definite. Finally, we use c to denote universal, estimated constants.

2.1 Graphical Submanifolds

Lemma 1 *Let $u : \mathbb{R}^n \to \mathbb{R}$ be smooth. Then* graph u *is a submanifold in $\mathbb{R}^{n+1}$. The metric g_{ij}, the lower unit normal vector ν, the second fundamental form h_{ij}, the mean curvature H, and the Gauß curvature K are given by*

$$\begin{aligned}
g_{ij} &= \delta_{ij} + u_i u_j,\\
g^{ij} &= \delta^{ij} - \frac{u^i u^j}{1+|Du|^2},\\
\nu &= \frac{((u_i), -1)}{\sqrt{1+|Du|^2}} \equiv \frac{((u_i), -1)}{\mathrm{v}},\\
h_{ij} &= \frac{u_{ij}}{\sqrt{1+|Du|^2}} \equiv \frac{u_{ij}}{\mathrm{v}},\\
H &= \operatorname{div}\left(\frac{Du}{\sqrt{1+|Du|^2}}\right),
\end{aligned}$$

and

$$K = \frac{\det D^2 u}{\left(1+|Du|^2\right)^{\frac{n+2}{2}}},$$

where $u_i \equiv \frac{\partial u}{\partial x^i}$, $u^i = u_j \delta^{ji}$ and $u_{ij} = \frac{\partial^2 u}{\partial x^i \partial x^j}$. Note that in Euclidean space, we often do not distinguish between Du and ∇u.

Proof

(i) We use the embedding vector $X(x) := (x, u(x))$, $X : \mathbb{R}^n \to \mathbb{R}^{n+1}$. The induced metric is the pull-back of the Euclidean metric in $\mathbb{R}^{n+1}$, $g := X^* g_{\mathbb{R}^{n+1}_{\text{Eucl.}}}$. We have $X_{,i} = (e_i, u_i)$. Hence

$$g_{ij} = X^{\alpha}_{,i}\delta_{\alpha\beta}X^{\beta}_{,j} = \langle X_{,i}, X_{,j}\rangle = \langle (e_i, u_i), (e_j, u_j)\rangle = \delta_{ij} + u_i u_j.$$

(ii) It is easy to check, that g^{ij} is the inverse of g_{ij}. Note that $u^i := \delta^{ij} u_j$, i.e., we lift the index with respect to the flat metric.

(iii) The vectors $X_{,i} = (e_i, u_i)$ are tangent to graph u. The vector $((-u_i), 1) \equiv (-Du, 1)$ is orthogonal to these vectors, hence, up to normalization, a unit normal vector.

(iv) We combine (1), (2) and compute the scalar product with ν to get

$$\begin{aligned} h_{ij} &= -\langle X_{;ij}, \nu\rangle = -\langle X_{,ij} - \Gamma^k_{ij} X_{,k}, \nu\rangle = -\langle X_{,ij}, \nu\rangle \\ &= -\left\langle (0, u_{ij}), \frac{((u_i), -1)}{\mathrm{v}}\right\rangle = \frac{u_{ij}}{\mathrm{v}}. \end{aligned}$$

(v) We obtain

$$\begin{aligned} H = \sum_{i=1}^{n} \lambda_i = g^{ij} h_{ij} &= \left(\delta^{ij} - \frac{u^i u^j}{1+|Du|^2}\right) \frac{u_{ij}}{\sqrt{1+|Du|^2}} \\ &= \frac{\delta^{ij} u_{ij}}{\sqrt{1+|Du|^2}} - \frac{u^i u^j u_{ij}}{\left(1+|Du|^2\right)^{3/2}} \\ &= \frac{\Delta u}{\sqrt{1+|Du|^2}} - \frac{u^i u^j u_{ij}}{\left(1+|Du|^2\right)^{3/2}} \end{aligned}$$

and, on the other hand,

$$\begin{aligned} \operatorname{div}\left(\frac{Du}{\sqrt{1+|Du|^2}}\right) &= \sum_{i=1}^{n} \frac{\partial}{\partial x^i} \frac{u_i}{\sqrt{1+|Du|^2}} \\ &= \sum_{i=1}^{n} \frac{u_{ii}}{\sqrt{1+|Du|^2}} - \sum_{i,j=1}^{n} \frac{u_i u_j u_{ji}}{\left(1+|Du|^2\right)^{3/2}} \\ &= H. \end{aligned}$$

(vi) From the defining equation for the principal curvatures and $\det g_{ij} = 1+|Du|^2$, we obtain

$$K = \prod_{i=1}^{n} \lambda_i = \det \left(g^{ij}h_{jk}\right) = \det g^{ij} \cdot \det h_{ij} = \frac{\det h_{ij}}{\det g_{ij}}$$

$$= \frac{v^{-n} \det u_{ij}}{v^2} = \frac{\det D^2 u}{\left(1+|Du|^2\right)^{\frac{n+2}{2}}}.$$

□

These formulae extend to the situation, in which u is defined on an open subset of $\mathbb{R}^n$.

Exercise 1 (Spheres) The lower hemisphere of radius R is locally given as graph u with $u : B_R(0) \to \mathbb{R}$ defined by $u(x) := -\sqrt{R^2 - |x|^2}$. Compute all the quantities mentioned in Lemma 1 and the principal curvatures explicitly for this example.

Exercise 2 Give a geometric definition of the (principal) curvature of a curve in $\mathbb{R}^2$ in terms of a circle approximating that curve in an optimal way.

Use the min-max characterization of eigenvalues to extend that geometric definition to n-dimensional hypersurfaces in $\mathbb{R}^{n+1}$.

Exercise 3 (Rotationally Symmetric Graphs) Assume that the function $u : \mathbb{R}^n \to \mathbb{R}$ is smooth and $u(x) = u(y)$, if $|x| = |y|$. Then $u(x) = f(|x|)$ for some $f : \mathbb{R}_+ \to \mathbb{R}$. Compute once again all the geometric quantities mentioned in Lemma 1.

3 Evolving Submanifolds

3.1 *General Assumption*

We will only consider the evolution of manifolds of dimension n embedded into $\mathbb{R}^{n+1}$, i.e. the evolution of hypersurfaces in Euclidean space. (Mean curvature flow is also considered for manifolds of arbitrary codimension. Another generalisation is to study flow equations of hypersurfaces immersed into Riemannian or Lorentzian manifolds.)

Definition 1 Let M^n denote an orientable manifold of dimension n. Let $X(\cdot, t) : M^n \to \mathbb{R}^{n+1}$, $0 \le t \le T < \infty$, be a smooth family of smooth embeddings. Let ν denote one choice of the normal vector field along $X(M^n, t)$. Then X or $(M_t)_{0\le t<T}$ with $M_t := X(M^n, t)$ is said to move with normal velocity F, if

$$\frac{d}{dt}X = -F\nu \quad \text{in } M^n \times [0, T).$$

Remark 2 In codimension 1, we often do not need to assume that M^n is orientable: Let $X : M^n \to N^{n+1}$ be a C^2-immersion and $H_1(N; \mathbb{Z}/2\mathbb{Z}) = 0$. Assume that X is proper, $X^{-1}(\partial N) = \partial M$, and X is transverse to ∂N. Then $N \setminus X(M)$ is not connected [13]. Hence, if M^n is closed and embedded in $\mathbb{R}^{n+1}$, M^n is orientable.

3.2 Evolution of Graphs

Lemma 2 *Let* $u : \mathbb{R}^n \times [0, \infty) \to \mathbb{R}$ *be a smooth function such that* graph u *evolves according to* $\frac{d}{dt}X = -F\nu$. *Then*

$$\dot{u} = \sqrt{1 + |Du|^2} \cdot F.$$

This result also holds, if u is defined on an open subset of $\mathbb{R}^n \times [0, \infty)$.

Proof Beware of assuming that the $(n + 1)$-st component in the evolution equation $\frac{d}{dt}X = -F\nu$ were equal to $\dot{u}$ as a hypersurface evolving according to $\frac{d}{dt}X = -F\nu$ does not only move in vertical direction but also in horizontal direction.

Let p denote a point on the abstract manifold embedded via X into $\mathbb{R}^{n+1}$. As our embeddings are graphical, we see that

$$X(p, t) = (x(p, t), u(x(p, t), t)).$$

We consider the scalar product of both sides of the evolution equation with ν and obtain

$$F = \langle F\nu, \nu\rangle = \left\langle -\frac{d}{dt}X, \nu\right\rangle = -\left\langle \left(\left(\dot{x}^k\right), u_i\dot{x}^i + \dot{u}\right), \frac{((u_i), -1)}{\sqrt{1 + |Du|^2}}\right\rangle = \frac{\dot{u}}{\sqrt{1 + |Du|^2}}.$$

□

Corollary 1 *Let* $u : \mathbb{R}^n \times [0, \infty) \to \mathbb{R}$ *be a smooth function such that* graph u *solves mean curvature flow* $\frac{d}{dt}X = -H\nu$. *Then*

$$\dot{u} = \sqrt{1 + |Du|^2} \cdot \operatorname{div}\left(\frac{Du}{\sqrt{1 + |Du|^2}}\right).$$

Exercise 4 (Rotationally Symmetric Translating Solutions) Let $u := \mathbb{R}^n \times \mathbb{R} \to \mathbb{R}$ be rotationally symmetric. Assume that graph u is a translating solution to mean curvature flow $\frac{d}{dt}X = -H\nu$, i.e. a solution such that $\dot{u}$ is constant.

Similar to Exercise 3, derive an ordinary differential equation for translating rotationally symmetric solutions to mean curvature flow.

Why does it suffice to consider the case $\dot{u} = 1$?

Remark 3 Consider a physical system consisting of a domain $\Omega \subset \mathbb{R}^3$. Assume that the energy of the system is proportional to the surface area of $\partial\Omega$. Then, up to a transformation $t \mapsto \mu t$ for some $\mu > 0$, the L^2-gradient flow for the area is mean curvature flow. We check that in a model case for graphical solutions in Lemma 3.

Lemma 3 *Let $u : \mathbb{R}^n \times [0, \infty) \to \mathbb{R}$ be smooth. Assume that $u(x, t) \equiv 0$ for $|x| \geq R$. Then the surface area is maximally reduced among all normal velocities F with given L^2-norm, if the normal velocity of* graph u *is given by H, i.e. if $\dot{u} = \sqrt{1 + |Du|^2} \cdot H$.*

Note that in general, soluIons to $\dot{u} = \sqrt{1 + |Du|^2} \cdot H$ do not have compact support.

Proof The area of graph $u(\cdot, t)|_{B_R}$ is given by

$$A(t) = \int\limits_{B_R} \sqrt{1 + |Du|^2}\, dx.$$

Define the induced area element $d\mu$ by $d\mu := \sqrt{1 + |Du|^2}\, dx$. We obtain using integration by parts

$$\begin{aligned} \frac{d}{dt} A(t)\bigg|_{t=0} &= \int\limits_{B_R} \frac{d}{dt} \sqrt{1 + |Du|^2}\, dx \Bigg|_{t=0} = \int\limits_{B_R(0)} \frac{1}{\sqrt{1 + |Du|^2}} \langle Du, D\dot{u} \rangle \Bigg|_{t=0} \\ &= -\int\limits_{B_R} \operatorname{div}\left(\frac{Du}{\sqrt{1 + |Du|^2}} \right) \frac{\dot{u}}{\mathrm{v}} \cdot \mathrm{v}\, dx \Bigg|_{t=0} = -\int\limits_{B_R} H\, F\, d\mu \Bigg|_{t=0} \\ &\geq -\left(\int\limits_{B_R} H^2\, d\mu \right)^{1/2} \left(\int\limits_{B_R} F^2\, d\mu \right)^{1/2} \Bigg|_{t=0}. \end{aligned}$$

Here, we have used Hölder's inequality $\|ab\|_{L^1} \leq \|a\|_{L^2} \cdot \|b\|_{L^2}$. There, we get equality precisely if a and b differ only by a multiplicative constant. Hence the surface area is reduced most efficiently among all normal velocities F with $\|F\|_{L^2} = \|H\|_{L^2}$, if we choose $F = H$. In this sense, mean curvature flow is the L^2-gradient flow for the area integral. □

3.3 Examples

Lemma 4 *Consider mean curvature flow, i.e. the evolution equation* $\frac{d}{dt}X = -H\nu$, *with* $M_0 = \partial B_R(0)$. *Then a smooth solution exists for* $0 \le t < T := \frac{1}{2n}R^2$ *and is given by* $M_t = \partial B_{r(t)}(0)$ *with* $r(t) = \sqrt{2n(T-t)} = \sqrt{R^2 - 2nt}$.

Proof The mean curvature of a sphere of radius $r(t)$ is given by $H = \frac{n}{r(t)}$. Hence we obtain a solution to mean curvature flow, if $r(t)$ fulfills

$$\dot{r}(t) = \frac{-n}{r(t)}.$$

A solution to this ordinary differential equation is given by $r(t) = \sqrt{2n(T-t)}$.

(The theory of partial differential equations implies that this solution is actually unique and hence no solutions exist that are not spherical.) □

Exercise 5 Find a solution to mean curvature flow with $M_0 = \partial B_R(0) \times \mathbb{R}^k \subset \mathbb{R}^l \times \mathbb{R}^k$. This includes in particular cylinders. Note that for $k \ge 1$, it is not obvious, whether these solutions are unique.

Exercise 6 Find solutions for $\frac{d}{dt}X = -|A|^2\nu$, $\frac{d}{dt}X = -K\nu$, $\frac{d}{dt}X = \frac{1}{H}\nu$, and $\frac{d}{dt}X = \frac{1}{K}\nu$ if $M_0 = \partial B_R(0) \subset \mathbb{R}^{n+1}$, especially for $n = 2$.

Remark 4 (Level-Set Flow for $F > 0$*)* Let M_t be a family of smooth embedded hypersurfaces in $\mathbb{R}^{n+1}$ that move according to $\frac{d}{dt}X = -F\nu$ with $F > 0$. Impose the global assumption that each point $x \in \mathbb{R}^{n+1}$ belongs to at most one hypersurface M_t. Then we can (at least locally) define a function $u : \mathbb{R}^{n+1} \to \mathbb{R}$ by setting $u(x) = t$, if $x \in M_t$. That is, $u(x)$ is the time, at which the hypersurface passes through the point x. We differentiate the identity $t = u(X(p,t))$, use that for closed shrinking hypersurfaces, Du is a negative multiple of the outer unit normal ν and get

$$1 = \frac{d}{dt}u(X(p,t)) = \left\langle Du, \frac{d}{dt}X\right\rangle = \langle Du, -F\nu\rangle = F \cdot |Du|.$$

We obtain the equation $F \cdot |Du| = 1$.

If $F < 0$, Du is a positive multiple of ν and we get $F \cdot |Du| = -1$.

This formulation is used to describe weak solutions, where singularities in the classical formulation occur. See for example [21], where the inverse mean curvature flow $F = -\frac{1}{H}$ is considered to prove the Riemannian Penrose inequality. Note that $H = \operatorname{div}\left(\frac{Du}{|Du|}\right)$ as the outer unit normal vector to a closed expanding hypersurface $M_t = \{u = t\}$ is given by $\frac{Du}{|Du|}$. According to (3), the divergence of the unit normal yields the mean curvature as the derivative of the unit normal in the direction of the unit normal vanishes. Hence the evolution equation $\frac{d}{dt}X = \frac{1}{H}\nu$ can be rewritten as

$$\operatorname{div}\left(\frac{Du}{|Du|}\right) = |Du|.$$

For contracting hypersurfaces under mean curvature flow with $H > 0$, the outer unit normal is given by $-\frac{Du}{|Du|}$ and $H = -\operatorname{div}\left(\frac{Du}{|Du|}\right)$. Hence mean curvature flow can be rewritten as $|Du| \cdot \operatorname{div}\left(\frac{Du}{|Du|}\right) = -1$.

Exercise 7 Verify the formula for the mean curvature in the level-set formulation. Compute level-set solutions to the flow equations $\frac{d}{dt}X = -H\nu$ and $\frac{d}{dt}X = \frac{1}{H}\nu$, where u depends only on $|x|$, i.e. the hypersurfaces M_t are spheres centered at the origin. Compare the result to your earlier computations.

We will use the level-set formulation to study a less trivial solution to mean curvature flow which can be written down in closed form.

Exercise 8 (Paper-Clip Solution) Let $\mathrm{v} \neq 0$. Consider the set

$$M_t := \left\{(x, y) \in \mathbb{R}^2 : e^{\mathrm{v}^2 t} \cosh(\mathrm{v}y) = \cos(\mathrm{v}x)\right\}.$$

Show that M_t solves mean curvature flow. Describe the shape of M_t for $t \to -\infty$ and for $t \nearrow 0$ (after appropriate rescaling).

Compare this to Theorem 3.

Note that you may also rewrite solutions equivalently (on an appropriate domain) as

$$y_\pm := \frac{1}{\mathrm{v}} \log\left(\cos(\mathrm{v}x) \pm \sqrt{\cos^2(\mathrm{v}x) - e^{2\mathrm{v}^2 t}}\right) - \mathrm{v}t.$$

Hint: You should obtain $t_x = u_x = -\frac{\sin(\mathrm{v}x)}{\mathrm{v}\cos(\mathrm{v}x)}$ and $u_y = -\frac{\sinh(\mathrm{v}y)}{\mathrm{v}\cosh(\mathrm{v}y)}$.

Remark 5 (Level-Set Flow) If a hypersurface moves with velocity F, where F is not necessarily positive, we cannot use the level-set formulation from above. Instead, we can use a function $u : \mathbb{R}^n \times [0, \infty) \to \mathbb{R}$ such that for each $c \in \mathbb{R}$, the set $M_t := \{x \in \mathbb{R}^n : u(x, t) = c\}$ (if it is a smooth hypersurface) is an embedded hypersurface that moves with velocity F.

We fix the unit normal $\nu = \frac{Du}{|Du|}$. Recall that $\dot{X} = -F\nu$. If u is as described above, we have $u(X(p, t), t) = c$ along the flow. Differentiating this equation yields $0 = \dot{u} + Du \cdot \dot{X} = \dot{u} + Du \cdot (-\nu) \cdot F = \dot{u} - |Du| \cdot F$.

For mean curvature flow, we obtain

$$\dot{u} = |Du| \cdot \operatorname{div}\left(\frac{Du}{|Du|}\right) = \left(\delta^{ij} - \frac{u^i u^j}{|Du|^2}\right) u_{ij}.$$

We leave it as an exercise that the converse implication is also true if the level sets are regular in the sense that $Du \neq 0$, i.e. that $\{x : u(x,t) = c\}$ evolves with normal velocity F if $\dot{u} = |Du| \cdot F$ and $Du \neq 0$ along $\{x : u(x,t) = c\}$.

3.4 *Short-Time Existence and Avoidance Principle*

In the case of closed initial hypersurfaces, short-time existence is guaranteed by the following

Theorem 1 (Short-Time Existence) *Let $X_0 : M^n \to \mathbb{R}^{n+1}$ be an embedding describing a smooth closed hypersurface. Let $F = F(\lambda_i)$ be smooth, symmetric, and $\frac{\partial F}{\partial \lambda_i} > 0$ everywhere on $X(M^n)$ for all i. Then the initial value problem*

$$\begin{cases} \frac{d}{dt}X = -F\nu, \\ X(\cdot, 0) = X_0 \end{cases}$$

has a smooth solution on some (short) time interval $[0, T)$, $T > 0$.

Proof (Idea of Proof) Represent potential solutions locally as graphs in a tubular neighbourhood of $X_0(M^n)$. Then $\frac{\partial F}{\partial \lambda_i} > 0$ ensures that the evolution equation for the height function in this coordinate system is strictly parabolic. Linear theory and the implicit function theorem guarantee that there exists a solution on a short time interval.

For more details see [22, Theorem 3.1]. □

Exercise 9

(i) Check, for which initial data the conditions in Theorem 1 are fulfilled if $F = H,\ K,\ |A|^2,\ -1/H,\ -1/K$.
(ii) Find examples of closed hypersurfaces such that

a) $H > 0$,
b) $K > 0$,
c) H is not positive everywhere,
d) $H > 0$, but K changes sign.

(iii) Show that on every smooth closed hypersurface $M^n \subset \mathbb{R}^{n+1}$, there is a point, where M^n is strictly convex, i.e. $\lambda_i > 0$ is fulfilled for every i.

On the other hand, starting with a closed hypersurface gives rise to solutions that exist at most on a finite time interval. This is a consequence of the avoidance principle. We will only consider the avoidance principle for mean curvature flow:

Theorem 2 (Avoidance Principle) *Let M_t^1 and $M_t^2 \subset \mathbb{R}^{n+1}$ be two embedded closed hypersurfaces and smooth solutions to $\frac{d}{dt}X = -H\nu$ on a common time interval $[0, T)$. If M_0^1 and M_0^2 are disjoint, then M_t^1 and M_t^2 are also disjoint.*

In particular, if M_0^1 is contained in a bounded component of $\mathbb{R}^{n+1} \setminus M_0^2$, then M_t^1 is contained in a bounded component of $\mathbb{R}^{n+1} \setminus M_t^2$.

Proof Suppose not. Then there would be some minimal $t_0 > 0$ such that $M_{t_0}^2$ touches $M_{t_0}^1$ at some point $p \in \mathbb{R}^{n+1}$. We get for the normals $\nu^1 = \pm\nu^2$ at p. Observe that if we change ν to $-\nu$, H also changes sign and $H\nu$ remains unchanged. Therefore it does not matter for mean curvature flow, which normal we choose and we may assume without loss of generality that $\nu^1 = \nu^2$ at p. Writing M_t^i locally as graph u^i over the common tangent hyperplane $T_pM_{t_0}^i \subset \mathbb{R}^{n+1}$, we see that the functions u^i fulfill

$$\dot{u}^i = \sqrt{1+|Du^i|^2} \cdot \operatorname{div}\left(\frac{Du^i}{\sqrt{1+|Du^i|^2}}\right) \equiv F\left(D^2u^i, Du^i\right).$$

We may assume that $u^1 > u^2$ for $t < t_0$. The evolution equation for the difference $w := u^1 - u^2$ fulfills $w > 0$ for $t < t_0$ locally in space-time and $w(0, t_0) = 0$, if we have $p = (0, 0)$ in our coordinate system. The evolution equation for w can be computed as follows

$$\begin{aligned}
\dot{w} &= \dot{u}^1 - \dot{u}^2 = F\left(D^2u^1, Du^1\right) - F\left(D^2u^2, Du^2\right)\\
&= \int_0^1 \frac{d}{d\tau} F\left(\tau D^2u^1 + (1-\tau)D^2u^2, \tau Du^1 + (1-\tau)Du^2\right) d\tau\\
&= \int_0^1 \frac{\partial F}{\partial r_{ij}}(\ldots)\, d\tau \cdot \left(u^1 - u^2\right)_{ij} + \int_0^1 \frac{\partial F}{\partial p_i}(\ldots)\, d\tau \cdot \left(u^1 - u^2\right)_i\\
&\equiv a^{ij} w_{ij} + b^i w_i.
\end{aligned}$$

Hence we can apply the parabolic Harnack inequality or the strong parabolic maximum principle and see that it is impossible that $w(x, t) > 0$ for small $|x|$ and $t < t_0$, but $w(0, t_0) = 0$. Hence M_t^1 cannot touch M_t^2 in a point, where $\nu^1 = \nu^2$. The theorem follows. □

Remark 6 The avoidance principle also extends to other normal velocities.

However, if $F\nu$ is not invariant under changing ν to $-\nu$, we have to ensure that the normals do not point in opposite directions, e.g. by assuming that one hypersurface encloses the other initially.

Usually, the normal velocity F, considered as a function of the principal curvatures, is defined on a convex cone $\Gamma \subset \mathbb{R}^n$. However, this does not ensure in general that F, considered as a function of $\left(D^2u, Du\right)$, is also defined on a convex set. Therefore we recommend in those cases to interpolate between the principal curvatures instead.

Exercise 10 Show that the normal velocities as considered in Exercise 9 can be represented (in an appropriate domain) as smooth functions of $\left(D^2u, Du\right)$ for hypersurfaces that are locally represented as graph u.

Corollary 2 (Finite Existence Time) *Let M_0 be a smooth closed embedded hypersurface in $\mathbb{R}^{n+1}$. Then a smooth solution M_t to $\frac{d}{dt}X = -H\nu$ can only exist on some finite time interval $[0, T)$, $T < \infty$.*

Proof Choose a large sphere that encloses M_0. According to Lemma 4, that sphere shrinks to a point in finite time. Thus the solution M_t can exist smoothly at most up to that time. □

Exercise 11 Deduce similar corollaries for the normal velocities in Exercise 9. You may use Exercise 6.

Remark 7 (Maximal Existence Time) Consider T maximal such that a smooth solution M_t as in Corollary 2 exists on $[0, T)$. Then the embedding vector X is uniformly bounded according to Theorem 2. Then some spatial derivative of the embedding $X(\cdot, t)$ has to become unbounded as $t \nearrow T$. For otherwise we could apply Arzelà-Ascoli and obtain a smooth limiting hypersurface M_T such that M_t converges smoothly to M_T as $t \nearrow T$. This, however, is impossibly, as Theorem 1 would allow to restart the flow from M_T. In this way, we could extend the flow smoothly all the way up to $T + \varepsilon$ for some $\varepsilon > 0$, contradicting the maximality of T.

It can often be shown that extending a solution beyond T is possible provided that $\|X(\cdot, t)\|_{C^2}$ is uniformly bounded. For mean curvature flow, this follows from explicit estimates. For other normal velocities, additional assumptions (the principal curvatures stay in a region, where F has nice properties) and Krylov-Safonov-estimates may be used to show such a result.

4 Evolution Equations for Submanifolds

In this chapter, we will compute evolution equations of geometric quantities, see e.g. [20, 22, 27].

For a family M_t of hypersurfaces solving the evolution equation

$$\frac{d}{dt}X = -F\nu \tag{6}$$

with $F = F(\lambda_i)$, where F is a smooth symmetric function, we have the following evolution equations.

Lemma 5 *The metric g_{ij} evolves according to*

$$\frac{d}{dt} g_{ij} = -2Fh_{ij}. \tag{7}$$

Proof By definition, $g_{ij} = \langle X_{,i}, X_{,j} \rangle = X^\alpha_{,i} \delta_{\alpha\beta} X^\beta_{,j}$. We differentiate with respect to time. Derivatives of $\delta_{\alpha\beta}$ vanish. The term $X^\alpha_{,i}$ involves only partial derivatives. We obtain

$$\frac{d}{dt} g_{ij} = \left(\dot{X}^\alpha\right)_{,i} \delta_{\alpha\beta} X^\beta_{,j} + X^\alpha_{,i} \delta_{\alpha\beta} \left(\dot{X}^\beta\right)_{,j}$$

(we may exchange partial spatial and time derivatives)

$$= \left(-F\nu^\alpha\right)_{,i} \delta_{\alpha\beta} X^\beta_{,j} + X^\alpha_{,i} \delta_{\alpha\beta} \left(-F\nu^\beta\right)_{,j}$$

(in view of the evolution equation $\frac{d}{dt} X = -F\nu$)

$$= -\, F\nu^\alpha_{;i} \delta_{\alpha\beta} X^\beta_{,j} - X^\alpha_{,i} \delta_{\alpha\beta} F\nu_{;j}$$

(terms involving derivatives of F vanish as ν and $X^\alpha_{,i}$ are orthogonal to each other; as the background metric $\overline{g}_{\alpha\beta} = \delta_{\alpha\beta}$ is flat, covariant and partial derivatives of ν coincide)

$$= -\, Fh^k_i X^\alpha_{,k} \delta_{\alpha\beta} X^\beta_{,j} - FX^\alpha_{,i} \delta_{\alpha\beta} h^k_j X^\beta_{,k}$$

(in view of the Weingarten equation (3))

$$= -\, Fh^k_i g_{kj} - Fg_{ik} h^k_j$$

(by the definition of the metric)

$$= -\, 2Fh_{ij}$$

(by the definition of $h^i_j := h_{jk} g^{ki}$).

The lemma follows. □

Corollary 3 *The evolution equation of the volume element* $d\mu := \sqrt{\det g_{ij}}\,dx$ *is given by*

$$\frac{d}{dt}d\mu = -FH\,d\mu. \tag{8}$$

Proof Exercise. Recall the formula for differentiating the determinant. □

Lemma 6 *The unit normal* ν *evolves according to*

$$\frac{d}{dt}\nu^\alpha = g^{ij}F_{;i}X^\alpha_{;j}. \tag{9}$$

Proof By definition, the unit normal vector ν has length one,

$$\langle \nu, \nu\rangle = 1 = \nu^\alpha \delta_{\alpha\beta}\nu^\beta.$$

Differentiating yields

$$0 = \dot{\nu}^\alpha \delta_{\alpha\beta}\nu^\beta.$$

Hence it suffices to show that the claimed equation is true if we take on both sides the scalar product with an arbitrary tangent vector. The vectors $X_{,i}$ (which we will also denote henceforth by X_i as there is no danger of confusion; we will also use this convention in other situations if partial and covariant derivatives of some quantity coincide) form a basis of the tangent plane at a fixed point. We differentiate the relation

$$0 = \langle \nu, X_i\rangle = \nu^\alpha \delta_{\alpha\beta}X_i^\beta$$

and obtain

$$\begin{aligned}
0 &= \frac{d}{dt}\nu^\alpha \delta_{\alpha\beta}X_i^\beta + \nu^\alpha \delta_{\alpha\beta}\frac{d}{dt}X_i^\beta\\
&= \frac{d}{dt}\nu^\alpha \delta_{\alpha\beta}X_i^\beta + \nu^\alpha \delta_{\alpha\beta}\left(\frac{d}{dt}X^\beta\right)_i\\
&= \frac{d}{dt}\nu^\alpha \delta_{\alpha\beta}X_i^\beta - \nu^\alpha \delta_{\alpha\beta}\left(F\nu^\beta\right)_i.
\end{aligned}$$

Hence

$$\begin{aligned}
\frac{d}{dt}\nu^\alpha \delta_{\alpha\beta}X_i^\beta &= \nu^\alpha \delta_{\alpha\beta}\nu^\beta F_i + F\nu^\alpha \delta_{\alpha\beta}\nu_i^\beta\\
&= F_i + F\tfrac{1}{2}\langle \nu, \nu\rangle_i = F_i
\end{aligned}$$

and the lemma follows as taking the scalar product of the claimed evolution equation with X_k, i.e. multiplying it with $\delta_{\alpha\beta}X_k^\beta$, yields

$$\frac{d}{dt}\nu^\alpha\delta_{\alpha\beta}X_k^\beta = g^{ij}F_iX_j^\alpha\delta_{\alpha\beta}X_k^\beta = g^{ij}F_ig_{jk} = \delta_k^iF_i = F_k.$$

□

Lemma 7 *The second fundamental form h_{ij} evolves according to*

$$\frac{d}{dt}h_{ij} = F_{;\,ij} - Fh_i^kh_{kj}. \tag{10}$$

Proof The Gauß formula (2) implies that $h_{ij} = -X_{;ij}^\alpha\nu_\alpha$. Differentiating yields

$$\begin{aligned}
\frac{d}{dt}h_{ij} &= -\frac{d}{dt}\langle X_{;ij}, \nu\rangle \\
&= -\left\langle\frac{d}{dt}X_{;ij}, \nu\right\rangle - \left\langle -h_{ij}\nu, \frac{d}{dt}\nu\right\rangle \\
&= -\left\langle\frac{d}{dt}X_{;ij}, \nu\right\rangle + h_{ij}\left\langle \nu, \frac{d}{dt}\nu\right\rangle \\
&= -\left\langle\frac{d}{dt}X_{;ij}, \nu\right\rangle \\
&= -\frac{d}{dt}\left(X_{,ij}^\alpha - \Gamma_{ij}^kX_k^\alpha\right)\nu_\alpha \\
&= -\left(\frac{d}{dt}X^\alpha\right)_{,ij}\nu_\alpha + \Gamma_{ij}^k\left(\frac{d}{dt}X^\alpha\right)_{,k}\nu_\alpha
\end{aligned}$$

(where no time derivatives of Γ_{ij}^k show up as $X_k^\alpha\nu_\alpha = 0$)

$$= (F\nu^\alpha)_{,ij}\nu_\alpha - \Gamma_{ij}^k(F\nu^\alpha)_{,k}\nu_\alpha$$

(in view of the evolution equation)

$$\begin{aligned}
&= F_{,ij}\nu^\alpha\nu_\alpha + F_{,i}\nu_{,j}^\alpha\nu_\alpha + F_{,j}\nu_{,i}^\alpha\nu_\alpha + F\nu_{,ij}^\alpha\nu_\alpha - \Gamma_{ij}^kF_{,k}\nu^\alpha\nu_\alpha - \Gamma_{ij}^kF\nu_{,k}^\alpha\nu_\alpha \\
&= F_{;ij} + F\nu_{,ij}^\alpha\nu_\alpha
\end{aligned}$$

as $F_{;ij} = F_{,ij} - \Gamma^k_{ij} F_{,k}$ and $\nu^\alpha_{,j} \nu_\alpha = \frac{1}{2}(\nu^\alpha \nu_\alpha)_j = 0$. It remains to show that $\nu^\alpha_{,ij} \nu_\alpha = -h^k_i h_{kj}$. We obtain

$$\nu^\alpha_{,ij} \nu_\alpha = \nu^\alpha_{;i,j} \nu_\alpha$$

(as $\nu^\alpha_i = \nu^\alpha_{;i}$)

$$= \nu^\alpha_{;ij} \nu_\alpha$$

($\nu^\alpha_{;ij} = (\nu^\alpha_{;i})_{,j} - \Gamma^k_{ij} \nu^\alpha_k$ and $0 = \nu^\alpha_k \nu_\alpha$)

$$= \left(h^k_i X^\alpha_k\right)_{;j} \nu_\alpha$$

(according to the Weingarten equation (3))

$$= h^k_i (-h_{kj} \nu^\alpha) \nu_\alpha$$

(due to the Gauß equation (2) and the orthogonality $X^\alpha_k \nu_\alpha = 0$)

$$= -\, h^k_i h_{kj}$$

as claimed. The Lemma follows. □

Lemma 8 *The normal velocity F evolves according to*

$$\frac{d}{dt} F - F^{ij} F_{;ij} = F F^{ij} h^k_i h_{kj}. \tag{11}$$

Proof We have, see [26, Lemma 5.4], the proof of [17, Theorem 2.1.20], or check this explicitly for the normal velocity considered,

$$\frac{\partial F}{\partial g_{kl}} = -F^{il} h^k_i$$

and compute the evolution equation of the normal velocity F

$$\begin{aligned} \frac{d}{dt} F - F^{ij} F_{;ij} = &- F^{il} h^k_i \frac{d}{dt} g_{kl} + F^{ij} \frac{d}{dt} h_{ij} - F^{ij} F_{;\,ij} \\ = &F F^{ij} h^k_i h_{kj}, \end{aligned}$$

where we used (7) and (10). □

We will need more explicit evolution equations for geometric quantities $\boxplus$ involving $\frac{d}{dt}\boxplus - F^{ij}\boxplus_{;ij}$.

Lemma 9 *The second fundamental form h_{ij} evolves according to*

$$\begin{aligned}\frac{d}{dt}h_{ij} - F^{kl}h_{ij;\,kl} = {}& F^{kl}h_k^a h_{al}\cdot h_{ij} - F^{kl}h_{kl}\cdot h_i^a h_{aj}\\ & - Fh_i^k h_{kj} + F^{kl,\,rs}h_{kl;\,i}h_{rs;\,j}.\end{aligned} \tag{12}$$

Proof Direct calculations yield

$$\begin{aligned}\frac{d}{dt}h_{ij} - F^{kl}h_{ij;\,kl} &= F_{;ij} - Fh_i^k h_{kj} - F^{kl}h_{ij;kl} && \text{by (10)}\\
&= F^{kl}h_{kl;\,ij} + F^{kl,\,rs}h_{kl;\,i}h_{rs;\,j}\\
&\quad - Fh_i^k h_{kj} - F^{kl}h_{ij;\,kl}\\
&= F^{kl}h_{ik;\,lj} + F^{kl,\,rs}h_{kl;\,i}h_{rs;\,j}\\
&\quad - Fh_i^k h_{kj} - F^{kl}h_{ik;\,jl} && \text{by Codazzi}\\
&= F^{kl}\left(h_k^a R_{ailj} + h_i^a R_{aklj}\right) - Fh_i^k h_{kj}\\
&\quad + F^{kl,\,rs}h_{kl;\,i}h_{rs;\,j} && \text{by (5)}\\
&= F^{kl}h_k^a h_{al}h_{ij} - F^{kl}h_k^a h_{aj}h_{il}\\
&\quad + F^{kl}h_i^a h_{al}h_{kj} - F^{kl}h_i^a h_{aj}h_{kl}\\
&\quad - Fh_i^k h_{kj} + F^{kl,\,rs}h_{kl;\,i}h_{rs;\,j} && \text{by (4)}\\
&= F^{kl}h_k^a h_{al}h_{ij} - F^{kl}h_i^a h_{aj}h_{kl}\\
&\quad - Fh_i^k h_{kj} + F^{kl,\,rs}h_{kl;\,i}h_{rs;\,j}.\end{aligned}$$

□

Remark 8 A direct consequence of (6) and (2) is

$$\frac{d}{dt}X^\alpha - F^{ij}X^\alpha_{;ij} = \left(F^{ij}h_{ij} - F\right)\nu^\alpha. \tag{13}$$

Hence

$$\frac{d}{dt}|X|^2 - F^{ij}\left(|X|^2\right)_{;ij} = 2\left(F^{ij}h_{ij} - F\right)\langle X,\,\nu\rangle - 2F^{ij}g_{ij}.$$

Proof We have

$$\begin{aligned}\frac{d}{dt}|X|^2 - F^{ij}\left(|X|^2\right)_{;ij} &= 2\left\langle X, \frac{d}{dt}X\right\rangle - 2F^{ij}\langle X_i, X_j\rangle - 2F^{ij}\langle X, X_{;ij}\rangle\\ &= 2\langle X, -F\nu\rangle - 2F^{ij}g_{ij} - 2F^{ij}\langle X, -h_{ij}\nu\rangle.\end{aligned}$$

□

Lemma 10 *The evolution equation for the unit normal* ν *is*

$$\frac{d}{dt}\nu^\alpha - F^{ij}\nu^\alpha_{;ij} = F^{ij}h^k_i h_{kj}\cdot\nu^\alpha. \tag{14}$$

Proof We compute

$$\begin{aligned}\frac{d}{dt}\nu^\alpha - F^{ij}\nu^\alpha_{;ij} &= g^{ij}F_{;\,i}X^\alpha_{;\,j} - F^{ij}\left(h^k_i X^\alpha_{;\,k}\right)_{;\,j} && \text{by (9) and (3)}\\ &= g^{ij}F^{kl}h_{kl;\,i}X^\alpha_{;\,j} - F^{ij}h^k_{i;\,j}X^\alpha_{;\,k} - F^{ij}h^k_i X^\alpha_{;\,kj}\\ &= F^{ij}h^k_i h_{kj}\nu^\alpha && \text{by (2).}\end{aligned}$$

□

Lemma 11 *The evolution equation for the scalar product* $\langle X, \nu\rangle$ *is*

$$\frac{d}{dt}\langle X, \nu\rangle - F^{ij}\langle X, \nu\rangle_{;ij} = -F^{ij}h_{ij} - F + F^{ij}h^k_i h_{kj}\langle X, \nu\rangle. \tag{15}$$

Proof We obtain

$$\begin{aligned}\frac{d}{dt}\langle X, \nu\rangle - F^{ij}\langle X, \nu\rangle_{;ij} &= X^\alpha\delta_{\alpha\beta}\left(\frac{d}{dt}\nu^\beta - F^{ij}\nu^\alpha_{;ij}\right)\\ &\quad + \left(\frac{d}{dt}X^\alpha - F^{ij}X^\alpha_{;\,ij}\right)\delta_{\alpha\beta}\nu^\beta\\ &\quad - 2F^{ij}X^\alpha_{;\,i}\delta_{\alpha\beta}\nu^\beta_{;\,j}\\ &= F^{ij}h^k_i h_{kj}\langle X, \nu\rangle + \left(F^{ij}h_{ij} - F\right)\langle \nu, \nu\rangle\\ &\quad - 2F^{ij}X^\alpha_{;\,i}\delta_{\alpha\beta}h^k_j X^\beta_{;\,k}\end{aligned}$$

by (3), (13), and (14)

$$= F^{ij}h^k_i h_{kj}\langle X, \nu\rangle - F^{ij}h_{ij} - F.$$

□

Lemma 12 *Let* $(\eta_\alpha)_\alpha = -e_{n+1} = (0, \ldots, 0, -1)$. *Then* $\tilde{\text{v}} := \langle \eta, \nu \rangle \equiv \eta_\alpha \nu^\alpha$ *fulfills*

$$\frac{d}{dt}\tilde{\text{v}} - F^{ij}\tilde{\text{v}}_{;ij} = F^{ij}h_i^k h_{kj}\tilde{\text{v}} \tag{16}$$

and $\text{v} := \tilde{\text{v}}^{-1}$ *fulfills*

$$\frac{d}{dt}\text{v} - F^{ij}\text{v}_{;ij} = -\text{v}F^{ij}h_i^k h_{kj} - 2\frac{1}{\text{v}}F^{ij}\text{v}_i\text{v}_j. \tag{17}$$

Proof The evolution equation for $\tilde{\text{v}}$ is a direct consequence of (14). For the proof of the evolution equation of v observe that

$$\text{v}_i = -\tilde{\text{v}}^{-2}\tilde{\text{v}}_i = -\text{v}^2\tilde{\text{v}}_i$$

and

$$\text{v}_{;ij} = -\tilde{\text{v}}^{-2}\tilde{\text{v}}_{;ij} + 2\tilde{\text{v}}^{-3}\tilde{\text{v}}_i\tilde{\text{v}}_j = -\text{v}^2\tilde{\text{v}}_{;ij} + 2\text{v}^{-1}\text{v}_i\text{v}_j.$$

□

5 Convex Hypersurfaces

5.1 Mean Curvature Flow

G. Huisken obtained the following theorem [20] for $n \geq 2$. The corresponding result for curves by M. Gage, R. Hamilton, and M. Grayson is even better, see [15, 18]. It is only required that $M \subset \mathbb{R}^2$ is a closed embedded curve.

Theorem 3 *Let* $M \subset \mathbb{R}^{n+1}$, $n \geq 2$, *be a smooth closed convex hypersurface. Then there exists a smooth family* M_t *of hypersurfaces solving*

$$\begin{cases} \frac{d}{dt}X = -H\nu & \text{for } 0 \leq t < T, \\ M_0 = M \end{cases}$$

for some $T > 0$.

As $t \nearrow T$,

- $M_t \to Q$ *in Hausdorff distance for some* $Q \in \mathbb{R}^{n+1}$ *(convergence to a point),*
- $(M_t - Q) \cdot (2n(T - t))^{-1/2} \to \mathbb{S}^n$ *smoothly (convergence to a "round point").*

The key step in the proof of Theorem 3 (in the case $n \geq 2$) is the following:

Theorem 4 *Let $M_t \subset \mathbb{R}^{n+1}$, $n \geq 2$, be a family of convex closed hypersurfaces flowing according to mean curvature flow. Then there exists some $\delta > 0$ such that*

$$\max_{M_t} \frac{n|A|^2 - H^2}{H^{2-\delta}}$$

is bounded above.

The proof involves complicated integral estimates.

Exercise 12 Prove Theorem 4 for $\delta = 0$.
Hint: Use Kato's inequality.

Theorem 5 (Kato's Inequality) *We have*

$$|\nabla|A||^2 \leq |\nabla A|^2.$$

Proof We prove this inequality if $|A| \neq 0$. In the exercise above, we only need that case. As $\nabla|A|^2 = 2|A|\nabla|A|$, the claim is equivalent to $\frac{1}{4}\left|\nabla|A|^2\right|^2 \leq |A|^2 \cdot |\nabla A|^2$. We choose a coordinate system such that in a fixed point $g_{ij} = \delta_{ij}$ and h_{ij} is diagonal with eigenvalues λ_i. We obtain there

$$\begin{aligned}
\frac{1}{4}\left|\nabla|A|^2\right|^2 &= \frac{1}{4}\sum_k \left(\nabla_k|A|^2\right)^2 = \sum_{i,j,k} \lambda_i h_{ii;k} \lambda_j h_{jj;k} \\
&\leq \sum_{i,j,k} \left(\frac{1}{2}\lambda_i^2 h_{jj;k}^2 + \frac{1}{2}\lambda_j^2 h_{ii;k}^2\right) = \sum_{i,j,k} \lambda_i^2 h_{jj;k}^2 \leq \sum_{i,j,k,l} h_{ij;k}^2 \lambda_l^2 \\
&= |A|^2 \cdot |\nabla A|^2.
\end{aligned}$$

□

Remark 9 For simplicity, we will illustrate the significance of the quantity considered in Theorem 4 only in the case $n = 2$. These considerations extend to higher dimensions.

As

$$\begin{aligned}
2|A|^2 - H^2 &= 2(\lambda_1^2 + \lambda_2^2) - (\lambda_1 + \lambda_2)^2 \\
&= 2\lambda_1^2 + 2\lambda_2^2 - \lambda_1^2 - 2\lambda_1\lambda_2 - \lambda_2^2 \\
&= \lambda_1^2 - 2\lambda_1\lambda_2 + \lambda_2^2 \\
&= (\lambda_1 - \lambda_2)^2,
\end{aligned}$$

it measures the difference from being umbilic ($\lambda_1 = \lambda_2$) and vanishes precisely if M_t is a sphere. Here, we have used that, according to Codazzi, $\lambda_1 = \lambda_2$ everywhere implies that M_t is locally part of a sphere or hyperplane.

Assume that $\min_{M_t} H \to \infty$ as $t \nearrow T$. Assume also that $\lambda_1 \le \lambda_2$ and that the surfaces stay strictly convex, i.e. $\min_{M_t} \lambda_1 > 0$. Then Theorem 4 implies for any ε that there exists t_ε, such that for $t_\varepsilon \le t < T$

$$\varepsilon \ge \max_{M_t} H^{-\delta} \ge \frac{2|A|^2 - H^2}{H^2} = \frac{(\lambda_1 - \lambda_2)^2}{(\lambda_1 + \lambda_2)^2} \ge \frac{(\lambda_1 - \lambda_2)^2}{4\lambda_2^2} = \frac{1}{4}\left(\frac{\lambda_1}{\lambda_2} - 1\right)^2.$$

Hence $\frac{\lambda_1}{\lambda_2} \approx 1$ and thus this implies that M_t is, in terms of the principal curvatures λ_i, close to a sphere.

5.2 *Gauß Curvature Flow and Other Normal Velocities*

There are many results showing that convex hypersurfaces converge to round points under certain flow equations, see e.g. [1, 2, 6, 14–16, 23, 27, 28, 32].

Let us consider normal velocities of homogeneity bigger than one. In this case, the calculations, that lead to a theorem corresponding to Theorem 4 for mean curvature flow, are much simpler and rely only on the maximum principle.

Theorem 6 ([2, Proposition 3]) *Let M_t be a smooth family of closed strictly convex solutions to Gauß curvature flow $\frac{d}{dt}X = -K\nu$. Then*

$$t \mapsto \max_{M_t} (\lambda_1 - \lambda_2)^2$$

is non-increasing.

Proof Recall that $H^2 - 4K = (\lambda_1 + \lambda_2)^2 - 4\lambda_1\lambda_2 = (\lambda_1 - \lambda_2)^2 =: w$. For Gauß curvature flow, we have, according to Appendix 2,

$$F^{ij} = K^{ij} = \frac{\partial}{\partial h_{ij}} \frac{\det h_{kl}}{\det g_{kl}} = \frac{\det h_{kl}}{\det g_{kl}} \tilde{h}^{ij} = K\tilde{h}^{ij},$$
$$F^{ij,kl} = K\tilde{h}^{ij}\tilde{h}^{kl} - K\tilde{h}^{ik}\tilde{h}^{lj},$$

where $\left(\tilde{h}^{ij}\right)_{i,j}$ is the inverse of $(h_{ij})_{i,j}$. Recall the evolution equations (7), (11), and (12) which become for Gauß curvature flow

$$\frac{d}{dt} g_{ij} = -2Kh_{ij},$$
$$\frac{d}{dt} K - K\tilde{h}^{kl} K_{kl} = KK\tilde{h}^{ij} h_i^k h_{kj} = K^2 H,$$

and

$$\begin{aligned}\frac{d}{dt}h_{ij} - K\tilde{h}^{kl}h_{ij;kl} =& K\tilde{h}^{kl}h_k^a h_{al}h_{ij} - K\tilde{h}^{kl}h_{kl}h_i^a h_{aj} - Kh_i^k h_{kj} \\ &+ K\left(\tilde{h}^{kl}\tilde{h}^{rs} - \tilde{h}^{kr}\tilde{h}^{sl}\right)h_{kl;i}h_{rs;j} \\ =& KHh_{ij} - (n+1)Kh_i^a h_{aj} + K\left(\tilde{h}^{kl}\tilde{h}^{rs} - \tilde{h}^{kr}\tilde{h}^{sl}\right)h_{kl;i}h_{rs;j},\end{aligned}$$

where $n = 2$. We have

$$\begin{aligned}\frac{d}{dt}H - K\tilde{h}^{ij}H_{;ij} =& -h_{ij}g^{ik}g^{jl}\frac{d}{dt}g_{kl} + g^{ij}\left(\frac{d}{dt}h_{ij} - K\tilde{h}^{kl}h_{ij;kl}\right) \\ =& 2K|A|^2 + KH^2 - 3K|A|^2 + Kg^{ij}\left(\tilde{h}^{kl}\tilde{h}^{rs} - \tilde{h}^{kr}\tilde{h}^{sl}\right)h_{kl;i}h_{rs;j} \\ =& K\left(H^2 - |A|^2\right) + Kg^{ij}\left(\tilde{h}^{kl}\tilde{h}^{rs} - \tilde{h}^{kr}\tilde{h}^{sl}\right)h_{kl;i}h_{rs;j} \\ =& 2K^2 + Kg^{ij}\left(\tilde{h}^{kl}\tilde{h}^{rs} - \tilde{h}^{kr}\tilde{h}^{sl}\right)h_{kl;i}h_{rs;j},\end{aligned}$$

hence

$$\begin{aligned}\frac{d}{dt}w - K\tilde{h}^{ij}w_{;ij} =& 2H\left(\frac{d}{dt}H - K\tilde{h}^{ij}H_{;ij}\right) - 2K\tilde{h}^{ij}H_iH_j \\ &- 4\left(\frac{d}{dt}K - K\tilde{h}^{ij}K_{;ij}\right) \\ =& 2H\left(2K^2 + Kg^{ij}\left(\tilde{h}^{kl}\tilde{h}^{rs} - \tilde{h}^{kr}\tilde{h}^{sl}\right)h_{kl;i}h_{rs;j}\right) \\ &- 2K\tilde{h}^{ij}H_iH_j - 4K^2H \\ =& 2HKg^{ij}\left(\tilde{h}^{kl}\tilde{h}^{rs} - \tilde{h}^{kr}\tilde{h}^{sl}\right)h_{kl;i}h_{rs;j} - 2K\tilde{h}^{ij}H_iH_j.\end{aligned}$$

In a coordinate system, such that $g_{ij} = \delta_{ij}$ and $h_{ij} = \operatorname{diag}(\lambda_1, \lambda_2)$ in a fixed point, we obtain

$$\begin{aligned}\frac{d}{dt}w - K\tilde{h}^{ij}w_{;ij} =& 2KH\sum_{i,j,k=1}^{2}\frac{1}{\lambda_i\lambda_j}h_{ii;k}h_{jj;k} - 2KH\sum_{i,j,k=1}^{2}\frac{1}{\lambda_i\lambda_j}h_{ij;k}^2 \\ &- 2K\sum_{i,j,k=1}^{2}\frac{1}{\lambda_k}h_{ii;k}h_{jj;k}\end{aligned}$$

$$
\begin{aligned}
&= 2KH \sum_{\substack{i,j,k=1\\ i\neq j}}^{2} \frac{1}{\lambda_i\lambda_j} h_{ii;k}h_{jj;k} - 2KH \sum_{\substack{i,j,k=1\\ i\neq j}}^{2} \frac{1}{\lambda_i\lambda_j} h_{ij;k}^2 - 2K \sum_{i,j,k=1}^{2} \frac{1}{\lambda_k} h_{ii;k}h_{jj;k}\\
&= \frac{4KH}{\lambda_1\lambda_2}\left(h_{11;1}h_{22;1} - h_{12;1}^2 + h_{11;2}h_{22;2} - h_{12;2}^2\right)\\
&\quad - \frac{2K}{\lambda_1}(h_{11;1}+h_{22;1})^2 - \frac{2K}{\lambda_2}(h_{11;2}+h_{22;2})^2.
\end{aligned}
$$

From now on, we consider a positive spatial maximum of H^2-4K. There, we get $2Hg^{ij}h_{ij;k} - 4K\tilde{h}^{ij}h_{ij;k} = 0$ for $k=1,2$. In a coordinate system as above, this (divided by 2) becomes

$$
\begin{aligned}
0 &= Hh_{11;k} + Hh_{22;k} - 2\frac{K}{\lambda_1}h_{11;k} - 2\frac{K}{\lambda_2}h_{22;k}\\
&= (\lambda_1+\lambda_2-2\lambda_2)h_{11;k} + (\lambda_1+\lambda_2-2\lambda_1)h_{22;k}\\
&= (\lambda_1-\lambda_2)(h_{11;k}-h_{22;k}).
\end{aligned}
$$

This enables us to replace $h_{11;2}$ in the evolution equation in a positive critical point by $h_{22;2}$: $h_{11;2}=h_{22;2}$ and $h_{22;1}=h_{11;1}$. Using also the Codazzi equations, we can rewrite the evolution equation in a positive critical point as

$$
\begin{aligned}
\frac{d}{dt}w - K\tilde{h}^{ij}w_{;ij} &= 4(\lambda_1+\lambda_2)\left(h_{11;1}^2 - h_{22;2}^2 + h_{22;2}^2 - h_{11;1}^2\right)\\
&\quad - \frac{2K}{\lambda_1}(h_{11;1}+h_{22;1})^2 - \frac{2K}{\lambda_2}(h_{11;2}+h_{22;2})^2\\
&\le 0.
\end{aligned}
$$

Hence, by the parabolic maximum principle, see Theorem 18 for a version on a domain, the claim follows. □

A consequence of Theorem 6 is the following result, see [2, Theorem 1].

Theorem 7 *Let $M \subset \mathbb{R}^3$ be a smooth closed strictly convex surface. Then there exists a smooth family of closed strictly convex hypersurfaces solving Gauß curvature flow $\frac{d}{dt}X = -K\nu$ for $0 \le t < T$. As $t \nearrow T$, M_t converges to a round point.*

Proof (Sketch of Proof) The main steps are

(i) The convergence to a point is due to K. Tso [31]. There, the problem is rewritten in terms of the support function and considered in all dimensions. It is shown that a positive lower bound on the Gauß curvature is preserved during the evolution. This ensures that the surfaces stay convex. The evolution equation of

$$\frac{K}{\langle X, \nu\rangle - \frac{1}{2}R}$$

is used to estimate K from above as long as the surface encloses $B_R(0)$. Then, using further estimates, a bound on the principal curvatures follows. Parabolic Krylov-Safonov estimates imply bounds on higher derivatives.

(ii) Theorem 6.

(iii) Show that M_t is between spheres of radius $r_+(t)$ and $r_-(t)$ and center $q(t)$ with $\frac{r_+(t)}{r_-(t)} \to 1$ as $t \nearrow T$.

(iv) Show that the quotient $\frac{K(p,t)}{K_{r(t)}}$ converges to 1 as $t \nearrow T$. Here

$$r(t) = (3(T-t))^{1/3}$$

is the radius of a sphere flowing according to Gauß curvature flow that becomes singular at $t = T$ and $K_{r(t)} = (3(T-t))^{-2/3}$ its Gauß curvature. This involves a Harnack inequality for the normal velocity.

(v) Show that $\frac{\lambda_i}{(3(T-t))^{-1/3}} \to 1$ as $t \nearrow T$.

(vi) Obtain uniform a priori estimates for a rescaled version of the flow and hence smooth convergence to a round sphere.

□

Theorem 7 has recently been generalised to higher dimensions by other methods, see [3, 4].

5.3 The Tensor Maximum Principle

Often, the tensor maximum principle can be used to deduce a priori bounds.

We see directly from the parabolic maximum principle for tensors that a positive lower bound on the principal curvatures is preserved for surfaces moving with normal velocity $|A|^2$.

Lemma 13 *For a smooth closed strictly convex surface M in $\mathbb{R}^3$, flowing according to $\frac{d}{dt}X = -|A|^2\nu$, the minimum of the principal curvatures is non-decreasing.*

Proof We have $F = |A|^2 = h_{ij}g^{jk}h_{kl}g^{li}$, $F^{ij} = 2g^{ia}h_{ab}g^{bj}$, and $F^{ij,kl} = 2g^{ik}g^{jl}$. Consider $M_{ij} = h_{ij} - \varepsilon g_{ij}$ with $\varepsilon > 0$ so small that M_{ij} is positive semi-definite for some time t_0. We wish to show that M_{ij} is positive semi-definite for $t > t_0$. Using (7) and (12), we obtain

$$\frac{d}{dt}h_{ij} - F^{kl}h_{ij;\,kl} = 2\operatorname{tr} A^3 h_{ij} - 3|A|^2 h_i^k h_{kj} + 2g^{kr}g^{ls}h_{kl;\,i}h_{rs;\,j}.$$

In the evolution equation for M_{ij}, we drop the positive definite terms involving derivatives of the second fundamental form

$$\frac{d}{dt}M_{ij} - F^{kl}M_{ij;\,kl} \geq 2\operatorname{tr} A^3 h_{ij} - 3|A|^2 h_i^k h_{kj} + 2\varepsilon|A|^2 h_{ij}.$$

Let ξ be a zero eigenvalue of M_{ij} with $|\xi| = 1$, $M_{ij}\xi^j = h_{ij}\xi^j - \varepsilon g_{ij}\xi^j = 0$. So we obtain in a point with $M_{ij} \geq 0$

$$\begin{aligned}\left(2\operatorname{tr} A^3 h_{ij} - 3|A|^2 h_i^k h_{kj} + 2\varepsilon|A|^2 h_{ij}\right)\xi^i\xi^j &= 2\varepsilon \operatorname{tr} A^3 - 3\varepsilon^2|A|^2 + 2\varepsilon^2|A|^2\\ &= 2\varepsilon \operatorname{tr} A^3 - \varepsilon^2|A|^2\\ &\geq 2\varepsilon^2|A|^2 - \varepsilon^2|A|^2 > 0\end{aligned}$$

and the maximum principle for tensors, Theorem 19, stated in the case of a differential equation $\frac{d}{dt}M_{ij} = \ldots$, extends to the case of a differential inequality $\frac{d}{dt}M_{ij} \geq \ldots$ and implies the result. □

Exercise 13 Show that under mean curvature flow of closed hypersurfaces, the following inequalities are preserved during the flow.

(i) $H \geq 0$, $H > 0$,
(ii) $h_{ij} \geq 0$,
(iii) $\varepsilon H g_{ij} \leq h_{ij} \leq \beta H g_{ij}$ for $0 < \varepsilon \leq \frac{1}{n} < \beta < 1$.

Such estimates exist also for other normal velocities.

5.4 *Two Dimensional Surfaces*

Theorem 8 ([27]) *Let M_t be a family of closed strictly convex hypersurfaces evolving according to $\frac{d}{dt}X = -|A|^2\nu$. Then*

$$t \mapsto \max_{M_t} \frac{(\lambda_1 + \lambda_2)(\lambda_1 - \lambda_2)^2}{\lambda_1\lambda_2}$$

is non-increasing.

Exercise 14

(i) Prove Theorem 8.
Hint: In a positive critical point of $w := \frac{(\lambda_1+\lambda_2)(\lambda_1-\lambda_2)^2}{\lambda_1\lambda_2}$, for $F = |A|^2$, the evolution equation of w is given by

$$\begin{aligned}\frac{d}{dt}w - F^{ij}w_{;ij} = & -4(\lambda_1-\lambda_2)^2\lambda_1\lambda_2 \\ & -2\frac{5\lambda_1^8-4\lambda_1^7\lambda_2+46\lambda_1^6\lambda_2^2+48\lambda_1^5\lambda_2^3+72\lambda_1^4\lambda_2^4}{\left(\lambda_1^2+\lambda_1\lambda_2+\lambda_2^2\right)^2\lambda_1^4}h_{11;1}^2 \\ & -2\frac{44\lambda_1^3\lambda_2^5+34\lambda_1^2\lambda_2^6+8\lambda_1\lambda_2^7+3\lambda_2^8}{\left(\lambda_1^2+\lambda_1\lambda_2+\lambda_2^2\right)^2\lambda_1^4}h_{11;1}^2 \\ & -2\frac{5\lambda_2^8-4\lambda_2^7\lambda_1+46\lambda_2^6\lambda_1^2+48\lambda_2^5\lambda_1^3+72\lambda_2^4\lambda_1^4}{\left(\lambda_2^2+\lambda_2\lambda_1+\lambda_1^2\right)^2\lambda_2^4}h_{22;2}^2 \\ & -2\frac{44\lambda_2^3\lambda_1^5+34\lambda_2^2\lambda_1^6+8\lambda_2\lambda_1^7+3\lambda_1^8}{\left(\lambda_2^2+\lambda_2\lambda_1+\lambda_1^2\right)^2\lambda_2^4}h_{22;2}^2.\end{aligned}$$

(This is a longer calculation.)

(ii) Show that the only closed strictly convex surfaces contracting self-similarly (by homotheties) under $\frac{d}{dt}X = -|A|^2\nu$, are round spheres. A surface M_t is said to evolve by homotheties, if for all t_1, t_2, there exists $\lambda \in \mathbb{R}$ such that $M_{t_1} = \lambda M_{t_2}$.

(iii) Show that for closed strictly convex initial data M, there exists some $c > 0$ such that $\frac{1}{c} \le \frac{\lambda_1}{\lambda_2} + \frac{\lambda_2}{\lambda_1} \le c$ for surfaces evolving according to $\frac{d}{dt}X = -|A|^2\nu$ for all $0 \le t < T$, where T is, as usual, the maximal existence time.

Similar results also exist for expanding surfaces

Theorem 9 ([28]) *Let M_t be a family of closed strictly convex hypersurfaces evolving according to $\frac{d}{dt}X = \frac{1}{K}\nu$. Then*

$$t \mapsto \max_{M_t} \frac{(\lambda_1-\lambda_2)^2}{\lambda_1^2\lambda_2^2}$$

is non-increasing.

Exercise 15 Prove Theorem 9 and deduce consequences similar to those in Exercise 14.

Hint: In a critical point of $w := \frac{(\lambda_1-\lambda_2)^2}{\lambda_1^2\lambda_2^2}$, the evolution equation of w reads

$$\frac{d}{dt}w - F^{ij}w_{;ij} = -2\frac{(\lambda_1+\lambda_2)(\lambda_1-\lambda_2)^2}{\lambda_1^3\lambda_2^3} - \frac{8}{\lambda_1^6\lambda_2}h_{11;\,1}^2 - \frac{8}{\lambda_1\lambda_2^6}h_{22;\,2}^2.$$

5.5 Calculations on a Computer Algebra System

For checking the monotonicity of

$$t \mapsto \max_{M_t} \frac{(\lambda_1+\lambda_2)(\lambda_1-\lambda_2)^2}{\lambda_1\lambda_2},$$

see Theorem 8, the calculations become quite long. In the following we describe how the calculations leading to this theorem can be done by a computer provided that you trust these machines.

(i) Rewrite $w = \frac{(\lambda_1+\lambda_2)(\lambda_1-\lambda_2)^2}{\lambda_1\lambda_2}$ in terms of H and K, H and K in terms of g_{ij} and h_{ij} and finally g_{ij} and h_{ij} as a function of Du and D^2u, provided that the surface is locally described as graph u.

(ii) Proceed similarly with the normal velocity $|A|^2 = F\left(Du, D^2u\right)$. Then u fulfills the partial differential equation

$$\dot{u} = \sqrt{1+|Du|^2} \cdot F\left(Du, D^2u\right) \equiv \mathrm{v}F.$$

(iii) Differentiating this equation yields

$$\dot{u}_k = \mathrm{v}F_{r_{ij}}u_{ijk} + \mathrm{v}F_{p_i}u_{ik} + \frac{u^i}{\mathrm{v}}Fu_{ik},$$

where we have used $F = F(p,r)$, and then dropping lower order terms suggests to consider the linearised operator

$$LW := \dot{W} - \mathrm{v}F_{r_{ij}}W_{ij},$$

where v and F are evaluated at $\left(Du, D^2u\right)$.

(iv) We would like to show that w is non-increasing. This follows from the maximum principle if we can show that $\frac{d}{dt}w - F^{ij}w_{;ij} \equiv \frac{d}{dt}w - \frac{\partial F}{\partial h_{ij}}w_{;ij} \le 0$ in a positive maximum of w. By the chain rule, we get

$$\frac{\partial F}{\partial r_{ij}} = \frac{\partial F}{\partial h_{kl}} \cdot \frac{\partial h_{kl}}{\partial r_{ij}} = \frac{\partial F}{\partial h_{ij}} \cdot \frac{1}{\mathrm{v}}.$$

(v) The considerations in the last paragraph do not depend on the coordinate system. We choose a coordinate system such that a positive maximum is attained at the origin and $Du(0) = 0$. We may assume in addition that $D^2u(0)$ is diagonal. At the origin, both factors that distinguish covariant and partial derivatives in $w_{;ij} = w_{,ij} - \Gamma_{ij}^k w_{,k}$ vanish. Hence it suffices to show that $Lw|_{x=0} \le 0$. This can be carried out with the help of a computer.

The algorithm in words:

1. Write $w = w\left(Du, D^2u\right)$ and $F = F\left(Du, D^2u\right)$.
2. Compute the following derivatives in terms of derivatives of u: $F_{r_{ij}}$, $\dot{w}$, w_i, w_{ij}.
3. Combine those derivatives and get $Lw =: N_1$ in terms of derivatives of u.
4. Use the relations obtained from differentiating $\dot{u} = \text{v}F$, $\dot{u}_k = (\text{v}F)_k$ and $\dot{u}_{kl} = (\text{v}F)_{kl}$ to remove any time derivative from N_1: Call the result N_2.
5. As w is positive and maximal at the point we want to consider, we can solve $w_k = 0$ for u_{11k} and u_{22k}. We use this to replace the terms u_{112} and u_{221} in N_2 and get N_3.
6. Assume that $Du(0) = 0$ and $D^2u(0) = \left(\begin{smallmatrix} a & 0 \\ 0 & b \end{smallmatrix}\right)$ in N_3 to get N_4.
7. N_4 consists of three terms:
$$N_4 = A + Bu_{111}^2 + Cu_{222}^2,$$
no terms involving $u_{111}u_{222}$ show up. Observe that A, B and C do only depend on a and b and that B and C are equal up to interchanging a and b.
8. It is easy to see that $A \le 0$ and $B \le 0$ for $a, b \ge 0$ in the situation of Theorem 8.
If it is not obvious, whether these inequalities hold, Sturm's algorithm [30] can be used to check the underlying polynomials for positivity.
9. Applying the steps above for different choices of w can be used to find monotone quantities, see [27, 28].

Two warnings:

- Do not use the simplifications valid at a single point, especially $Du = 0$, before differentiating.
- The computer might identify u_{12} and u_{21}. Take this into account when computing $F_{r_{12}}$.

Exercise 16 Prove Theorem 8 based on computer algebra calculations.

6 Mean Curvature Flow of Entire Graphs

For mean curvature flow of entire graphs, K. Ecker and G. Huisken proved the following existence theorem [11, Theorem 5.1].

Theorem 10 *Let $u_0 : \mathbb{R}^n \to \mathbb{R}$ be locally Lipschitz continuous. Then there exists a function $u \in C^\infty(\mathbb{R}^n \times (0, \infty)) \cap C^0(\mathbb{R}^n \times [0, \infty))$ solving*

$$\begin{cases} \dot{u} = \sqrt{1 + |Du|^2} \cdot \operatorname{div}\left(\dfrac{Du}{\sqrt{1 + |Du|^2}}\right) & \text{in } \mathbb{R}^n \times (0, \infty), \\ u(\cdot, 0) = u_0 & \text{in } \mathbb{R}^n. \end{cases}$$

The key ingredient in the existence proof is the following localised gradient estimate.

Theorem 11 *Let $u : B_R(0) \times [0, T] \to \mathbb{R}$ be a smooth solution to graphical mean curvature flow. Then*

$$\sqrt{1 + |Du|^2(0,t)} \le c(n) \sup_{B_R(0)} \sqrt{1 + |Du|^2(\cdot, 0)} \cdot \exp\left(c(n)\, R^{-2} \left(\operatorname*{osc}_{B_R(0)\times[0,T]} u \right)^2 \right).$$

We do not prove this Theorem in this course. However, if we additionally assume that $u(x, 0) \to \infty$ as $|x| \to \infty$, Theorem 16, that is much easier to prove, can be used instead of Theorem 11.

Theorem 10 has been extended to continuous initial data by J. Clutterbuck [7] and T. Colding and W. Minicozzi [9].

If u is initially close to a cone in an appropriate sense, graphical mean curvature flow converges, as $t \to \infty$, after appropriate rescaling, to a self-similarly expanding solution "coming out of a cone", see the papers by K. Ecker and G. Huisken [11] and N. Stavrou [29].

Stability of translating solutions to graphical mean curvature flow without rescaling is considered in [8].

7 Mean Curvature Flow Without Singularities

The material in this section is based on joint work with M. Sáez, see [25].

7.1 Intuition

Remark 10

(i) Long time existence for entire graphs was first shown by K. Ecker and G. Huisken [11], see Theorem 10.
(ii) We wish to study the evolution of complete graphs defined on subsets of Euclidean space $\mathbb{R}^{n+1}$. The additional dimension is related to Theorem 13.
(iii) We assume for the moment that such initial data have smooth solutions. Then the following figures should give some intuition about the behaviour of these solutions.

a) A rotationally symmetric solution defined on a ball: Fig. 1 on page 108 shows a rotationally symmetric graph in $\mathbb{R}^{n+2}$ defined on a ball in $\mathbb{R}^{n+1}$. A cylinder over the boundary of the ball encloses this graph. Asymptotically, these two hypersurfaces coincide as $x^{n+2} \to \infty$. Under mean curvature flow, the cylinder in $\mathbb{R}^{n+2}$ collapses to a line in finite time. The sphere

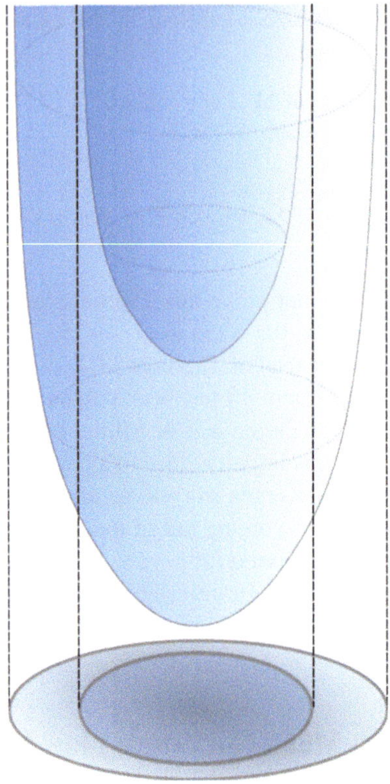

Fig. 1 Graph defined over a ball

in $\mathbb{R}^{n+1}$ collapses to a point in finite time. As the principal curvatures of any cylinder $M_t^n \times \mathbb{R}$ are $\lambda_1, \ldots, \lambda_n, 0$, where $\lambda_1, \ldots, \lambda_n$ are the principal curvatures of M_t^n, the projection of the evolving cylinder coincides at all times with the evolving sphere.

The evolution of the graph stays graphical and asymptotic to the evolving cylinder as $x^{n+2} \to \infty$. As the curvature near the tip is larger than that of the cylinder, the tip moves faster and moves up to infinity at precisely the time when the cylinder collapses to a line. Thus for all times, the boundary of the projections of the graphs coincides with the evolving spheres and hence fulfills mean curvature flow.

b) A solution initially defined on a domain that will form a neckpinch under mean curvature flow for $n \geq 2$: In Fig. 2 on page 109, the graph is initially defined over a domain whose boundary will develop a neckpinch in finite time, i.e. the thin neck will collapse. There are methods to continue the flow past this neckpinch singularity. After this singularity, the hypersurface splits into two topologically spherical components. Once again, the evolution of the graph above is such that the boundary of its projection or, equivalently, of the domain of definition of the graph, fulfills mean curvature flow. This happens as follows: As the neckpinch singularity forms downstairs, the

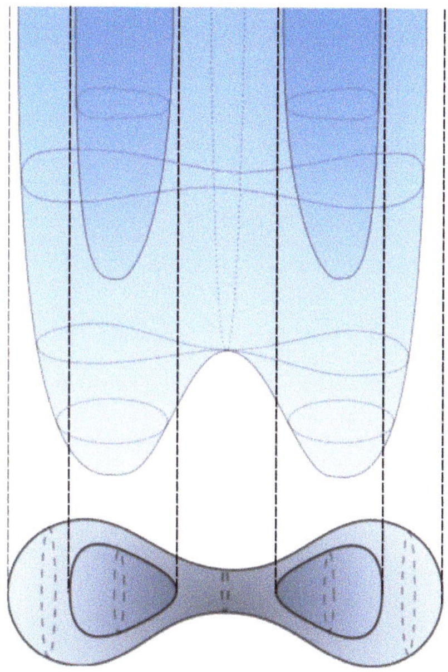

Fig. 2 Solution with a neckpinch singularity

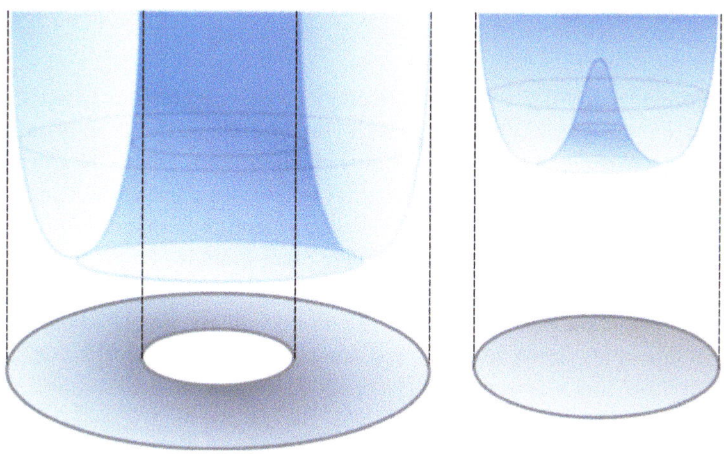

Fig. 3 Graph defined over an annulus

mean curvature in $\mathbb{R}^{n+1}$ blows up. Meanwhile, above the neck region in $\mathbb{R}^{n+2}$, the mean curvature becomes even larger so that the graph over the neck region moves to infinity while the rest of the graph remains finite. Then the graph separates into two disjoint components.

c) A solution initially defined on an annulus: In Fig. 3 on page 109, the domain of definition is an annulus. Its boundary consists of two disjoint spheres that

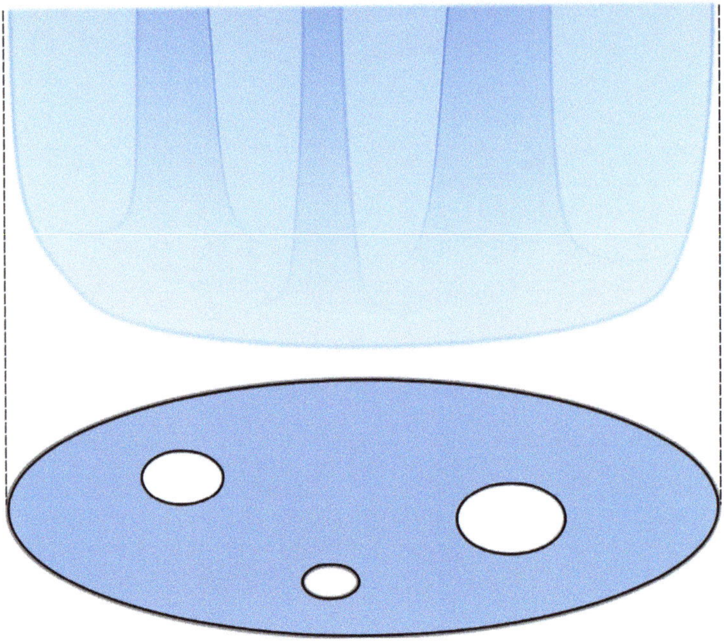

Fig. 4 Ω_t with many holes

disappear at different times. The graph above is asymptotic to two cylinders as $x^{n+2} \to \infty$. When the inner cylinder collapses, a "cap at infinity" is added to the graph and its topology changes. Similarly to the example of a contracting sphere, this cap can travel in finite time from infinity downwards and become visible. Later, the situation is similar to that of Fig. 1.

d) A solution defined on a domain in the plane bounded by possibly countably many disjoint curves: For a planar domain with finitely many holes, see Fig. 4 on page 110, there are finitely many times, where boundary components shrink to points and vanish similarly to the situation in Fig. 3. At those times, caps at infinity are added to the graphical solution similarly to the annulus situation above.

Finally, if a planar domain has countably many holes, we can arrange so that the holes disappear on a dense set of times. We get a smoothly evolving graph whose mean curvature is unbounded at all times.

7.2 Results

Let us consider mean curvature flow for graphs defined on a relatively open set

$$\Omega \equiv \bigcup_{t\geq 0} \Omega_t \times \{t\} \subset \mathbb{R}^{n+1} \times [0, \infty). \tag{18}$$

Our existence result for bounded domains is

Theorem 12 (Existence) *Let $A \subset \mathbb{R}^{n+1}$ be a bounded open set and $u_0 \colon A \to \mathbb{R}$ a locally Lipschitz continuous function with $u_0(x) \to \infty$ for $x \to x_0 \in \partial A$.*

Then there exists (Ω, u), where $\Omega \subset \mathbb{R}^{n+1} \times [0, \infty)$ is relatively open, such that $u \colon \Omega \to \mathbb{R}$ solves graphical mean curvature flow

$$\dot{u} = \sqrt{1 + |Du|^2} \cdot \operatorname{div} \left(\frac{Du}{\sqrt{1 + |Du|^2}} \right) \quad \text{in } \Omega \cap \{t > 0\},$$

u is smooth for $t > 0$ and continuous up to $t = 0$, $\Omega_0 = A$, $u(\cdot, 0) = u_0$ in A and $u(x, t) \to \infty$ as $(x, t) \to (x_0, t_0) \in \partial\Omega$, where $\partial\Omega$ is the relative boundary of Ω in $\mathbb{R}^{n+1} \times [0, \infty)$.

Such smooth solutions yield weak solutions to mean curvature flow. We have

Theorem 13 (Weak Flow) *Let (A, u_0) and (Ω, u) be as in Theorem 12. Let $\partial\mathcal{D}_t$ be the level set evolution of $\partial\Omega_0$ with $\mathcal{D}_0 = \Omega_0$. If $\partial\mathcal{D}_t$ does not fatten, the measure theoretic boundaries of Ω_t and $\mathcal{D}_t$ coincide for every $t \geq 0$.*

Here, $\mathcal{D}_t = \left\{x \in \mathbb{R}^{n+1} \colon w(x, t) < 0\right\}$ and w solves $\dot{w} = |Dw| \cdot \operatorname{div}\left(\frac{Dw}{|Dw|}\right)$ as in Remark 5. The equation is solved in the viscosity sense, see e.g. [5, 12] for more details.

7.3 Strategy of Proof

Proof (Strategy of the Proof of Theorem 12)

(i) Fix $L > 0$. Then there exists a solution with initial value $\min\{u_0, L\}$ for all $t \in [0, \infty]$, see [11].

(ii) If $L_1 < L$, we prove a priori estimates for the part of the evolving graphs which is below L_1. This is done in Theorem 16 for the (spatial) first order derivatives of u. See Theorem 17 for the second derivative bounds. Similar techniques imply bounds for all higher derivatives.

(iii) We let $L \to \infty$ and use a variant of the Theorem of Arzelà-Ascoli to pass to a subsequence which converges to our solution.

□

Proof (Sketch of the Strategy of the Proof of Theorem 13) In the following sketch of a proof we try to give an idea of the argument without mentioning technical details, e.g. approximations or fattening. None of the steps works exactly as described below.

(i) The constructed solution graph $u(\cdot, t)$ corresponds to a level-set solution.
(ii) The level-set solution starting from $\partial A \times \mathbb{R}$ is an outer barrier to the graphical solution graph $u(\cdot, t)$. Observe that Ω_t is the projection of the evolving graph at time t to $\mathbb{R}^{n+1}$. Hence Ω_t is contained in the level-set evolution of A.
(iii) By shifting the level set solution downwards, we obtain convergence to the level set solution starting with the cylinder $\partial A \times \mathbb{R}$. This prevents graph $u(\cdot, t)$ from detaching near infinity from the evolution of the cylinder.

□

7.4 The A Priori Estimates

Recall the definition $\mathrm{v} = \sqrt{1 + |Du|^2}$, where we consider u as a function defined on some subset of $\mathbb{R}^{n+1} \times [0, \infty)$.

Let $\eta := (\eta_\alpha) = (0, \ldots, 0, 1)$. In the following, whenever quantities like v or $|A|^2$ are involved, we consider u and v as functions on the evolving hypersurfaces rather than as functions depending on $(x, t) \in \mathbb{R}^{n+1} \times [0, \infty)$, i.e. we consider $u := X^\alpha \eta_\alpha$ and $\mathrm{v} := -\langle \nu, \eta \rangle^{-1}$.

Theorem 14 *Let X be a solution to mean curvature flow. Then we have the following evolution equations*

$$\begin{aligned} \left(\tfrac{d}{dt} - \Delta\right) u &= 0, \\ \left(\tfrac{d}{dt} - \Delta\right) \mathrm{v} &= -|A|^2 \mathrm{v} - \tfrac{2}{\mathrm{v}} |\nabla \mathrm{v}|^2, \\ \left(\tfrac{d}{dt} - \Delta\right) |A|^2 &= -2|\nabla A|^2 + 2|A|^4, \\ \left(\tfrac{d}{dt} - \Delta\right) \mathcal{G} &\le -2k \cdot \mathcal{G}^2 - 2\varphi \mathrm{v}^{-3} \langle \nabla \mathrm{v}, \nabla \mathcal{G} \rangle, \end{aligned}$$

where $\mathcal{G} = \varphi |A|^2 \equiv \frac{\mathrm{v}^2}{1 - k\mathrm{v}^2} |A|^2$ and $k > 0$ is chosen so that $k\mathrm{v}^2 \le \frac{1}{2}$ in the domain considered.

Proof For mean curvature flow, we have $F^{ij} = g^{ij}$. This implies $F^{ij} h_{ij} = H$. In view of (13), we deduce $\left(\frac{d}{dt} - \Delta\right) X = 0$ and $\left(\frac{d}{dt} - \Delta\right) u = 0$.

For the evolution equation of $w := |A|^2$, we calculate

$$\begin{aligned} \left(\tfrac{d}{dt} - \Delta\right) g_{ij} &= -2H h_{ij}, && \text{see (7)}, \\ \left(\tfrac{d}{dt} - \Delta\right) h_{ij} &= |A|^2 h_{ij} - 2H h_i^a h_{aj}, && \text{see (12)}, \\ w &= g^{ik} h_{ij} g^{jl} h_{kl}, \end{aligned}$$

$$
\begin{aligned}
\dot{w} &= 2g^{ik}\dot{h}_{ij}g^{jl}h_{kl} - 2g^{ir}g^{sk}h_{ij}g^{jl}h_{kl}\dot{g}_{rs},\\
w_r &= 2g^{ik}h_{ij;r}g^{jl}h_{kl},\\
w_{rs} &= 2g^{ik}h_{ij;rs}g^{jl}h_{kl} + 2g^{ik}h_{ij;r}g^{jl}h_{kl;s},\\
\left(\tfrac{d}{dt} - \Delta\right)|A|^2 &= 2g^{ik}\left(|A|^2h_{ij} - 2Hh_i^a h_{aj}\right)g^{jl}h_{kl} + 4H \operatorname{tr} A^3 - 2|\nabla A|^2\\
&= 2|A|^4 - 2|\nabla A|^2.
\end{aligned}
$$

For the remaining claims see [10, 11]. □

Assumption 15 *For the proof of the a priori estimates, we will assume that*

$$u\colon \mathbb{R}^{n+1} \times [0, \infty) \to \mathbb{R}$$

is a smooth solution to mean curvature flow such that for any $T > 0$ there exists $R > 0$ such that for all $t \in [0, T]$

$$\{x : u(x, t) \le 0\} \subset B_R(0).$$

In order to be able to consider smooth solutions, a few extra constructions are necessary.

Theorem 16 (C^1-Estimates) *Let u be as in Assumption 15. Then*

$$\mathrm{v}(-u)^2 = \mathrm{v}u^2 \le \max_{\substack{t=0\\ \{u<0\}}} \mathrm{v}u^2$$

at points where $u < 0$.

Here and in the following, it is often possible to increase the exponent of $-u$.

Proof According to Theorem 14, $w := \mathrm{v}u^2$ fulfills

$$
\begin{aligned}
\dot{w} &= \dot{\mathrm{v}}u^2 + 2\mathrm{v}u\dot{u},\\
w_i &= \mathrm{v}_i u^2 + 2\mathrm{v}uu_i,\\
w_{ij} &= \mathrm{v}_{ij}u^2 + 2\mathrm{v}uu_{ij} + 2\mathrm{v}u_iu_j + 2u(\mathrm{v}_iu_j + \mathrm{v}_ju_i),\\
\left(\tfrac{d}{dt} - \Delta\right) w &= u^2\left(\tfrac{d}{dt} - \Delta\right)\mathrm{v} - 2\mathrm{v}|\nabla u|^2 - 4u\langle\nabla \mathrm{v}, \nabla u\rangle\\
&= u^2\left(-\mathrm{v}|A|^2 - \frac{2}{\mathrm{v}}|\nabla \mathrm{v}|^2\right) - 2\mathrm{v}|\nabla u|^2 - 4\left\langle\frac{u}{\sqrt{\mathrm{v}}}\nabla \mathrm{v}, \sqrt{\mathrm{v}}\nabla u\right\rangle\\
&\le -u^2\mathrm{v}|A|^2 \le 0.
\end{aligned}
$$

The estimate follows from the maximum principle applied to w in the domain where $u < 0$. □

Remark 11 We recommend thinking of Theorem 16 as an estimate for $\mathrm{v}(-u)^2$.

Corollary 4 *Let u be as in Assumption 15. Then*

$$\mathrm{v} \le \max_{\substack{t=0\\ \{u<0\}}} \mathrm{v}u^2$$

at points where $u \le -1$.

Exercise 17 Consider $\mathrm{v}(-u)$ to obtain similar C^1-estimates.

Remark 12 Corollaries similar to Corollary 4 also hold for the following a priori estimates for points with $u \le -\varepsilon < 0$ or $t \ge \varepsilon > 0$. We do not write them down explicitly.

In Theorem 16 and later, the result still holds if we replace every u by $u - h$ for any constant h.

Remark 13 For later use, we estimate derivatives of u and v,

$$|\nabla u|^2 = \eta_\alpha X_i^\alpha g^{ij} X_j^\beta \eta_\beta = \eta_\alpha \left(\delta^{\alpha\beta} - \nu^\alpha \nu^\beta\right) \eta_\beta = 1 - \mathrm{v}^{-2} \le 1$$

and, according to (3),

$$\begin{aligned} |\nabla \mathrm{v}|^2 &= \left(\left(-\eta_\alpha \nu^\alpha\right)^{-1}\right)_i g^{ij} \left(\left(-\eta_\beta \nu^\beta\right)^{-1}\right)_j = \mathrm{v}^4 \eta_\alpha X_k^\alpha h_i^k g^{ij} h_j^l X_l^\beta \eta_\beta \\ &\le \mathrm{v}^4 |A|^2 \le \mathrm{v}^2 \varphi |A|^2 = \mathrm{v}^2 \mathscr{G}. \end{aligned}$$

We therefore obtain

$$|\langle \nabla u, \nabla \mathrm{v}\rangle| \le |\nabla u| \cdot |\nabla \mathrm{v}| \le \mathrm{v}^2 |A| \le \mathrm{v}\sqrt{\mathscr{G}}.$$

Theorem 17 **(C^2-Estimates)** *Let u be as in Assumption 15.*

(i) Then there exist $\lambda > 0$, $c > 0$ *and* $k > 0$ *(the constant in* φ *and implicitly in* $\mathscr{G}$*), depending on the* C^1*-estimates, such that*

$$tu^4 \mathscr{G} + \lambda u^2 \mathrm{v}^2 \le ct + \sup_{\substack{t=0\\ \{u<0\}}} \lambda u^2 \mathrm{v}^2$$

at points where $u < 0$ *and* $0 < t \le 1$.

(ii) *Moreover, if u is in C^2 initially, we get C^2-estimates up to $t = 0$: Then there exists $c > 0$, depending only on the C^1-estimates, such that*

$$u^4\mathscr{G} \le ct + \sup_{\substack{t=0\\ \{u<0\}}} u^4\mathscr{G}$$

at points where $u < 0$.

Proof In order to prove both parts simultaneously, we set

$$w := (\mu t + (1-\mu))u^4\mathscr{G} + \lambda u^2 \mathrm{v}^2 \equiv \mu_t u^4\mathscr{G} + \lambda u^2\mathrm{v}^2.$$

If we set $\mu = 1$, we obtain $\mu_t = t$ and later the first claim, if $\mu = \lambda = 0$, we get $\mu_t = 1$ and deduce in the following the second claim. We calculate

$$\begin{aligned}
\dot w =& \mu u^4\mathscr{G} + 4\mu_t u^3\mathscr{G}\dot u + \mu_t u^4\dot{\mathscr{G}} + 2\lambda \mathrm{v}^2 u\dot u + 2\lambda u^2 \mathrm{v}\dot{\mathrm{v}},\\
w_i =& 4\mu_t u^3\mathscr{G}u_i + \mu_t u^4\mathscr{G}_i + 2\lambda\mathrm{v}^2 uu_i + 2\lambda u^2\mathrm{v}\mathrm{v}_i,\\
w_{ij} =& 4\mu_t u^3\mathscr{G}u_{ij} + \mu_t u^4\mathscr{G}_{ij} + 2\lambda\mathrm{v}^2uu_{ij} + 2\lambda u^2\mathrm{v}\mathrm{v}_{ij} + 12\mu_t u^2\mathscr{G}u_iu_j\\
&+ 4\mu_t u^3(\mathscr{G}_iu_j + \mathscr{G}_ju_i) + 2\lambda\mathrm{v}^2u_iu_j + 2\lambda u^2\mathrm{v}_i\mathrm{v}_j\\
&+ 4\lambda u\mathrm{v}(u_i\mathrm{v}_j + u_j\mathrm{v}_i),\\
\mu_t u^3\nabla\mathscr{G} =& \frac{1}{u}\nabla w - 4\mu_t u^2\mathscr{G}\nabla u - 2\lambda\mathrm{v}^2\nabla u - 2\lambda u\mathrm{v}\nabla\mathrm{v},\\
\left(\tfrac{d}{dt} - \Delta\right) w \le& \mu u^4\mathscr{G} + \mu_t u^4\left(-2k\cdot\mathscr{G}^2 - 2\varphi\mathrm{v}^{-3}\langle\nabla\mathrm{v}, \nabla\mathscr{G}\rangle\right)\\
&+ 2\lambda u^2\mathrm{v}\left(-|A|^2\mathrm{v} - \tfrac{2}{\mathrm{v}}|\nabla\mathrm{v}|^2\right) - 12\mu_t u^2\mathscr{G}|\nabla u|^2\\
&- 8\mu_t u^3\langle\nabla\mathscr{G}, \nabla u\rangle - 2\lambda\mathrm{v}^2|\nabla u|^2 - 2\lambda u^2|\nabla\mathrm{v}|^2 - 8\lambda u\mathrm{v}\langle\nabla u, \nabla\mathrm{v}\rangle.
\end{aligned}$$

In the following, we will use the notation $\langle\nabla w, b\rangle$ with a generic vector b. The constants c are allowed to depend on $\sup\{|u|: u < 0\}$ (which does not exceed its initial value) and the C^1-estimates. It may also depend on an upper bound for t, but we assume that $0 < t \le 1$ whenever t appears explicitly. I.e., we suppress dependence on already estimated quantities.

We estimate the terms involving $\nabla\mathscr{G}$ separately. Let $\varepsilon > 0$ be a constant. We fix its value below. Using Remark 13 for estimating terms, we get

$$\begin{aligned}
&-2\varphi\mu_t u^4\mathrm{v}^{-3}\langle\nabla\mathrm{v}, \nabla\mathscr{G}\rangle\\
&\quad= -2\frac{\varphi u}{\mathrm{v}^3}\left\langle\nabla\mathrm{v}, \frac{1}{u}\nabla w - 4\mu_t u^2\mathscr{G}\nabla u - 2\lambda\mathrm{v}^2\nabla u - 2\lambda u\mathrm{v}\nabla\mathrm{v}\right\rangle\\
&\quad\le \langle\nabla w, b\rangle + 8\mu_t\frac{\varphi|u|^3}{\mathrm{v}}\mathscr{G}|A| + 4\lambda\varphi\mathrm{v}|u||A| + 4\frac{\lambda\varphi u^2}{\mathrm{v}^2}|\nabla\mathrm{v}|^2
\end{aligned}$$

$$
\begin{aligned}
&= \langle \nabla w, b\rangle + 8\mu_t\varphi^2 \frac{|u|^3 \mathscr{G}^{3/2}}{\varphi^{3/2}} \frac{1}{\mathrm{v}} + 4\lambda\varphi \mathrm{v}|u||A| + \lambda u^2 |\nabla \mathrm{v}|^2 \cdot 4\frac{\varphi}{\mathrm{v}^2} \\
&\le \langle \nabla w, b\rangle + \varepsilon\mu_t u^4 \mathscr{G}^2 + \varepsilon\lambda u^2 \mathrm{v}^2 |A|^2 + \lambda u^2 |\nabla \mathrm{v}|^2 \cdot 4\frac{\varphi}{\mathrm{v}^2} \\
&\quad + c(\varepsilon, \lambda),
\end{aligned}
$$

$$
\begin{aligned}
-8\mu_t u^3 \langle \nabla \mathscr{G}, \nabla u\rangle & \\
&= -8\left\langle \nabla u, \frac{1}{u}\nabla w - 4\mu_t u^2 \mathscr{G}\nabla u - 2\lambda \mathrm{v}^2 \nabla u - 2\lambda u \mathrm{v} \nabla \mathrm{v}\right\rangle \\
&\le \langle \nabla w, b\rangle + 32\mu_t u^2 \mathscr{G} + 16\lambda \mathrm{v}^2 + 16\lambda |u| \mathrm{v}^3 |A| \\
&\le \langle \nabla w, b\rangle + \varepsilon\mu_t u^4 \mathscr{G}^2 + \varepsilon\lambda u^2 \mathrm{v}^2 |A|^2 + c(\varepsilon, \lambda).
\end{aligned}
$$

We obtain

$$
\begin{aligned}
\left(\tfrac{d}{dt} - \Delta\right) w \le{}& \mu u^4 \mathscr{G} + \mu_t u^4 \mathscr{G}^2 (-2k + 2\varepsilon) + \langle \nabla w, b\rangle \\
&+ \lambda u^2 \mathrm{v}^2 |A|^2 (-2 + 3\varepsilon) + \lambda u^2 |\nabla \mathrm{v}|^2 \left(4\tfrac{\varphi}{\mathrm{v}^2} - 6\right) + c(\varepsilon, \lambda).
\end{aligned}
$$

Let us assume that $k > 0$ is chosen so small that $k\mathrm{v}^2 \le \frac{1}{3}$ in $\{u < 0\}$. This implies $\varphi \le 2\mathrm{v}^2$. We may assume that $\lambda \ge 2u^2$ in $\{u < 0\}$ and get $\mu u^4 \mathscr{G} \le \frac{1}{2}\lambda u^2 \varphi |A|^2 \le \lambda u^2 \mathrm{v}^2 |A|^2$. We get

$$
4\frac{\varphi}{\mathrm{v}^2} - 6 = \frac{4}{1 - k\mathrm{v}^2} - 6 \le 0.
$$

Finally, fixing $\varepsilon > 0$ sufficiently small, we obtain

$$
\left(\tfrac{d}{dt} - \Delta\right) w \le \langle \nabla w, b\rangle + c.
$$

Now, both claims follow from the maximum principle. □

Appendix 1: Parabolic Maximum Principles

The following maximum principle is fairly standard. For non-compact, strict or other maximum principles, we refer to [11] or [24], respectively.

We will use $C^{2;1}$ for the space of functions that are two times continuously differentiable with respect to the space variables and once continuously differentiable with respect to the time variable.

Theorem 18 (Weak Parabolic Maximum Principle) *Let $\Omega \subset \mathbb{R}^n$ be open and bounded and $T > 0$. Let $a^{ij}, b^i \in L^\infty(\Omega \times [0, T])$. Let a^{ij} be strictly elliptic, i.e. $a^{ij}(x, t) > 0$ in the sense of matrices. Let $u \in C^{2;1}(\Omega \times [0, T)) \times C^0\left(\overline{\Omega} \times [0, T]\right)$ fulfill*

$$\dot{u} \le a^{ij}u_{ij} + b^i u_i \quad \text{in } \Omega \times (0, T).$$

Then we get for $(x, t) \in \Omega \times (0, T)$

$$u(x, t) \le \sup_{\mathscr{P}(\Omega\times(0,T))} u,$$

where $\mathscr{P}(\Omega \times (0, T)) := (\Omega \times \{0\}) \cup (\partial\Omega \times (0, T))$.

Proof

(i) Let us assume first that $\dot{u} < a^{ij}u_{ij} + b^i u_i$ in $\Omega \times (0, T)$. If there exists a point $(x_0, t_0) \in \Omega \times (0, T)$ such that $u(x_0, t_0) > \sup\limits_{\mathscr{P}(\Omega\times(0,T))} u$, we find $(x_1, t_1) \in \Omega \times (0, T)$ and t_1 minimal such that $u(x_1, t_1) = u(x_0, t_0)$. At (x_1, t_1), we have $\dot{u} \ge 0$, $u_i = 0$ for all $1 \le i \le n$, and $u_{ij} \le 0$ (in the sense of matrices). This, however, is impossible in view of the evolution equation.

(ii) Define for $0 < \varepsilon$ the function $\mathrm{v} := u - \varepsilon t$. It fulfills the differential inequality

$$\dot{\mathrm{v}} = \dot{u} - \varepsilon < \dot{u} \le a^{ij}u_{ij} + b^i u_i = a^{ij}\mathrm{v}_{ij} + b^i \mathrm{v}_i.$$

Hence, by the previous considerations,

$$u(x, t) - \varepsilon t = \mathrm{v}(x, t) \le \sup_{\mathscr{P}(\Omega\times(0,T))} \mathrm{v} = \sup_{\mathscr{P}(\Omega\times(0,T))} u - \varepsilon t$$

and the result follows as $\varepsilon \searrow 0$.

□

There is also a parabolic maximum principle for tensors, see [19, Theorem 9.1]. (See the AMS-Review for a small correction of the proof.)

A tensor N_{ij} depending smoothly on M_{ij} and g_{ij}, involving contractions with the metric, is said to fulfill the null-eigenvector condition, if $N_{ij}\mathrm{v}^i\mathrm{v}^j \ge 0$ for all null-eigenvectors v of M_{ij}.

Theorem 19 *Let $(M_{ij})_{i,j}$ be a tensor, defined on a closed Riemannian manifold $(M, g(t))$, fulfilling*

$$\frac{\partial}{\partial t} M_{ij} = \Delta M_{ij} + b^k \nabla_k M_{ij} + N_{ij}$$

on a time interval $[0, T)$, where b is a smooth vector field and N_{ij} fulfills the null-eigenvector condition. If $M_{ij} \ge 0$ at $t = 0$, then $M_{ij} \ge 0$ for $0 \le t < T$.

Appendix 2: Some Linear Algebra

Lemma 14 *We have*

$$\frac{\partial}{\partial a_{ij}}\det(a_{rs}) = \det(a_{rs})a^{ji},$$

if a_{ij} is invertible with inverse a^{ij}, i.e. if $a^{ij}a_{jk} = \delta^i_k$.

Proof It suffices to prove that the claimed equality holds when we multiply it with a_{ik} and sum over i. Hence, we have to show that

$$\frac{\partial}{\partial a_{ij}}\det(a_{rs})a_{ik} = \det(a_{rs})\delta^j_k.$$

We get

$$\frac{\partial}{\partial a_{ij}}\det(a_{rs}) = \det\begin{pmatrix} a_{1\,1} & \dots & a_{1\,j-1} & 0 & a_{1\,j+1} & \dots & a_{1\,n} \\ \vdots & & \vdots & \vdots & \vdots & & \vdots \\ a_{i-1\,1} & \dots & a_{i-1\,j-1} & 0 & a_{i-1\,j+1} & \dots & a_{i-1\,n} \\ 0 & \dots & 0 & 1 & 0 & \dots & 0 \\ a_{i+1\,1} & \dots & a_{i+1\,j-1} & 0 & a_{i+1\,j+1} & \dots & a_{i+1\,n} \\ \vdots & & \vdots & \vdots & \vdots & & \vdots \\ a_{n\,1} & \dots & a_{n\,j-1} & 0 & a_{n\,j+1} & \dots & a_{n\,n} \end{pmatrix}.$$

and thus

$$\begin{aligned} \frac{\partial}{\partial a_{ij}}\det(a_{rs})\cdot a_{ik} = & \det\begin{pmatrix} 0 & \dots & 0 & a_{1\,k} & 0 & \dots & 0 \\ a_{2\,1} & \dots & a_{2\,j-1} & 0 & a_{2\,j+1} & \dots & a_{2\,n} \\ \vdots & & \vdots & \vdots & \vdots & & \vdots \\ a_{n\,1} & \dots & a_{n\,j-1} & 0 & a_{n\,j+1} & \dots & a_{n\,n} \end{pmatrix} \\ & + \det\begin{pmatrix} a_{1\,1} & \dots & a_{1\,j-1} & 0 & a_{1\,j+1} & \dots & a_{1\,n} \\ 0 & \dots & 0 & a_{2\,k} & 0 & \dots & 0 \\ a_{3\,1} & \dots & a_{3\,j-1} & 0 & a_{3\,j+1} & \dots & a_{3\,n} \\ \vdots & & \vdots & \vdots & \vdots & & \vdots \\ a_{n\,1} & \dots & a_{n\,j-1} & 0 & a_{n\,j+1} & \dots & a_{n\,n} \end{pmatrix} \\ & + \dots \end{aligned}$$

$$
\begin{aligned}
&= \det \begin{pmatrix} a_{1\,1} & \dots & a_{1\,j-1} & a_{1\,k} & a_{1\,j+1} & \dots & a_{1\,n} \\ a_{2\,1} & \dots & a_{2\,j-1} & 0 & a_{2\,j+1} & \dots & a_{2\,n} \\ \vdots & & \vdots & \vdots & \vdots & & \vdots \\ a_{n\,1} & \dots & a_{n\,j-1} & 0 & a_{n\,j+1} & \dots & a_{n\,n} \end{pmatrix} \\
&\quad + \det \begin{pmatrix} a_{1\,1} & \dots & a_{1\,j-1} & 0 & a_{1\,j+1} & \dots & a_{1\,n} \\ a_{2\,1} & \dots & a_{2\,j-1} & a_{2\,k} & a_{2\,j+1} & \dots & a_{2\,n} \\ a_{3\,1} & \dots & a_{3\,j-1} & 0 & a_{3\,j+1} & \dots & a_{3\,n} \\ \vdots & & \vdots & \vdots & \vdots & & \vdots \\ a_{n\,1} & \dots & a_{n\,j-1} & 0 & a_{n\,j+1} & \dots & a_{n\,n} \end{pmatrix} \\
&\quad + \dots \\
&= \det \begin{pmatrix} a_{1\,1} & \dots & a_{1\,j-1} & a_{1\,k} & a_{1\,j+1} & \dots & a_{1\,n} \\ \vdots & & \vdots & \vdots & \vdots & & \vdots \\ a_{n\,1} & \dots & a_{n\,j-1} & a_{n\,k} & a_{n\,j+1} & \dots & a_{n\,n} \end{pmatrix} \\
&= \delta^j_k \det(a_{rs}).
\end{aligned}
$$

□

Lemma 15 *Let $a_{ij}(t)$ be differentiable in t with inverse $a^{ij}(t)$. Then*

$$\frac{d}{dt} a^{ij} = -a^{ik} a^{lj} \frac{d}{dt} a_{kl}.$$

Proof We have

$$a^{ik} a_{kj} = \delta^i_j.$$

There exists $\tilde{a}^{ij}$ such that

$$a_{ik} \tilde{a}^{kj} = \delta^j_i.$$

Then $a^{ij} = \tilde{a}^{ij}$, as

$$a^{ij} = a^{ik} \delta^j_k = a^{ik} \left(a_{kl} \tilde{a}^{lj} \right) = \left(a^{ik} a_{kl} \right) \tilde{a}^{lj} = \tilde{a}^{ij}.$$

We differentiate and obtain

$$0 = \frac{d}{dt} \delta^i_j = \frac{d}{dt} \left(a^{ik} a_{kj} \right) = \frac{d}{dt} a^{ik} a_{kj} + a^{ik} \frac{d}{dt} a_{kj}.$$

Hence

$$\frac{d}{dt}a^{il} = \frac{d}{dt}a^{ik}\delta_k^l = \frac{d}{dt}a^{ik}a_{kj}a^{jl} = -a^{ik}\frac{d}{dt}a_{kj}a^{jl}.$$

□

References

1. B. Andrews, Contraction of convex hypersurfaces in Euclidean space. Calc. Var. Partial Differ. Equ. **2**(2), 151–171 (1994)
2. B. Andrews, Gauss curvature flow: the fate of the rolling stones. Invent. Math. **138**(1), 151–161 (1999)
3. B. Andrews, P. Guan, L. Ni, Flow by powers of the Gauss curvature. Adv. Math. **299**, 174–201 (2016)
4. S. Brendle, K. Choi, P. Daskalopoulos, Asymptotic behavior of flows by powers of the Gaussian curvature. Acta Math. **219**, 1–16 (2017)
5. Y.G. Chen, Y. Giga, S. Goto, Uniqueness and existence of viscosity solutions of generalized mean curvature flow equations. J. Differ. Geom. **33**(3), 749–786 (1991)
6. B. Chow, Deforming convex hypersurfaces by the nth root of the Gaussian curvature. J. Differ. Geom. **22**(1), 117–138 (1985)
7. J. Clutterbuck, Parabolic equations with continuous initial data (2004). arXiv:math.AP/0504455
8. J. Clutterbuck, O.C. Schnürer, F. Schulze, Stability of translating solutions to mean curvature flow. Calc. Var. Partial Differ. Equ. **29**(3), 281–293 (2007)
9. T.H. Colding, W.P. Minicozzi II, Sharp estimates for mean curvature flow of graphs. J. Reine Angew. Math. **574**, 187–195 (2004)
10. K. Ecker, *Regularity Theory for Mean Curvature Flow*. Progress in Nonlinear Differential Equations and Their Applications, vol. 57 (Birkhäuser Boston Inc., Boston, 2004)
11. K. Ecker, G. Huisken, Interior estimates for hypersurfaces moving by mean curvature. Invent. Math. **105**(3), 547–569 (1991)
12. L.C. Evans, J. Spruck, Motion of level sets by mean curvature. I. J. Differ. Geom. **33**(3), 635–681 (1991)
13. M.E. Feighn, Separation properties of codimension-1 immersions. Topology **27**(3), 319–321 (1988)
14. W.J. Firey, Shapes of worn stones. Mathematika **21**, 1–11 (1974)
15. M. Gage, R.S. Hamilton, The heat equation shrinking convex plane curves. J. Differ. Geom. **23**(1), 69–96 (1986)
16. C. Gerhardt, Flow of nonconvex hypersurfaces into spheres. J. Differ. Geom. **32**(1), 299–314 (1990)
17. C. Gerhardt, *Curvature Problems*. Series in Geometry and Topology, vol. 39 (International Press, Somerville, 2006)
18. M.A. Grayson, The heat equation shrinks embedded plane curves to round points. J. Differ. Geom. **26**(2), 285–314 (1987)
19. R.S. Hamilton, Three-manifolds with positive Ricci curvature. J. Differ. Geom. **17**(2), 255–306 (1982)
20. G. Huisken, Flow by mean curvature of convex surfaces into spheres. J. Differ. Geom. **20**(1), 237–266 (1984)
21. G. Huisken, T. Ilmanen, The inverse mean curvature flow and the Riemannian Penrose inequality. J. Differ. Geom. **59**(3), 353–437 (2001)

22. G. Huisken, A. Polden, Geometric evolution equations for hypersurfaces, in *Calculus of Variations and Geometric Evolution Problems (Cetraro, 1996)*. Lecture Notes in Mathematics, vol. 1713 (Springer, Berlin, 1999), pp. 45–84
23. J.A. McCoy, The surface area preserving mean curvature flow. Asian J. Math. **7**(1), 7–30 (2003)
24. M.H. Protter, H.F. Weinberger, *Maximum Principles in Differential Equations* (Springer, New York, 1984). Corrected reprint of the 1967 original
25. M. Sáez Trumper, O.C. Schnürer, Mean curvature flow without singularities. J. Differ. Geom. **97**(3), 545–570 (2014)
26. O.C. Schnürer, The Dirichlet problem for Weingarten hypersurfaces in Lorentz manifolds. Math. Z. **242**(1), 159–181 (2002)
27. O.C. Schnürer, Surfaces contracting with speed $|A|^2$. J. Differ. Geom. **71**(3), 347–363 (2005)
28. O.C. Schnürer, Surfaces expanding by the inverse Gauß curvature flow. J. Reine Angew. Math. **600**, 117–134 (2006)
29. N. Stavrou, Selfsimilar solutions to the mean curvature flow. J. Reine Angew. Math. **499**, 189–198 (1998)
30. C. Sturm, Mémoire sur la résolution des équations numeriques. Bull. Sci. Math. Ferussac **11**, 419–422 (1829)
31. K. Tso, Deforming a hypersurface by its Gauss-Kronecker curvature. Commun. Pure Appl. Math. **38**(6), 867–882 (1985)
32. J.I.E. Urbas, An expansion of convex hypersurfaces. J. Differ. Geom. **33**(1), 91–125 (1991)

MIX
Papier aus verantwortungsvollen Quellen
Paper from responsible sources
FSC® C105338

If you have any concerns about our products,
you can contact us on
ProductSafety@springernature.com

In case Publisher is established outside the EU,
the EU authorized representative is:
Springer Nature Customer Service Center GmbH
Europaplatz 3, 69115 Heidelberg, Germany

Printed by Libri Plureos GmbH
in Hamburg, Germany